THE CLIMATE SOLUTION

THE CLIMATE SOLUTION

B. VINCENT

CONTENTS

First Printing, 2024

CHAPTER 1

Introduction to Global Warming

The Earth's climate is changing. Climatic changes have naturally occurred over the last four billion years, but the current climatic change is exceptional with regard to both its cause and its consequences. This time it is us humans who are causing these changes. However, we can also find the solution, remedying the deleterious effects of our behavior. Indeed, by acting collectively, (nearly) everyone can help to prevent and moderate these changes. This is the aim of this book, which is addressed to economists and bel canto enthusiasts, two already rather cosmopolitan populations, but also to everyone who is affected by this risk. Our earth is indeed a community, and the attention and actions of all of its members are necessary. Before describing the solutions, we must verify whether in fact a problem exists and recognize that it is a global problem, one that is more immediately perceptible from Tokyo or New York than from Tamanrasset.

Since the end of the last glaciation, the Earth's climate has been on a warming trend. What is exceptional are neither the changes nor the great speed of these changes, but rather the unilinearity and persistence of this trend. Two characteristics are remarkable. First, the Earth's climate had never previously undergone such powerful forcings as the industrial activities that were not confined to any geographical region and that were concentrated in time. Second, the last time that the Earth experienced such warm temperatures was 125,000 years ago, a period of melting of the great glacial ice. The Earth's systems have not been disrupted by our industrial activities to the point of posing any risk to the latter's future. Our collective survival is not threatened, but it may become more costly to make adjustments in a few decades. While the overall risk over the centuries is comparable, this does not mean that we should not worry since the world of the coming century and the world of the coming millennium are different.

Understanding the Science of Climate Change

In 2018, the people of the world were confronted with a one-two punch that underscored the meaning of climate change. First came the Intergovernmental Panel on Climate Change (IPCC) report, which laid out what would happen if the increase in global warming was limited to 1.5°C from pre-industrial levels compared to 2.0°C. The difference between the two temperatures seemed small, but that small difference could have significant effects. Reaching 1.5°C meant a modest increase in global temperatures above those of June 2018, whereas reaching 2.0°C would mean substantial additional warming. In many ways, the story of 1.5°C was a reminder that what might seem like small quantitative differences are in fact significant qualitative ones. The IPCC story was followed in November by the State of California's wildfire story - this despite an aggressive program over more than a decade to thin out and manage forests, removing excess flammable vegetation, and thus reducing the propensity for catastrophic fires. These wildfires raised the question of whether changes in the climate had effectively overwhelmed these efforts or had changed the natural processes that had previously sustained the forests.

Historical Context and Impacts

The increase in GHGs has been relentless since the start of the industrial revolution. Their growing presence has created an atmospheric blanket that has raised the average temperature of the planet about 1.3°F over the past century. Projections of continuing increases in GHGs suggest that average global temperature will rise by an additional 4.5°F to 9°F by the end of the century. This change in temperature has and will continue to have far-reaching and acutely tragic impacts, with less developed nations and the poorest ethnic and racial minorities in all nations being the most vulnerable and heavily impacted. According to the IPCC, changes in climate have already affected numerous physical and biological systems, and will alter ecosystem services, as well as aspects of agriculture, forestry, fisheries, human health, settlement, and refugees.

There is clear and compelling evidence that Earth's climate is changing, with over 800 recorded examples that include accelerating ice loss, a warming ocean, frequent and intense heavy precipitation, and extended periods of oppressively high difficulty. The strength of the signal is now so high that it can be discerned from any location on the planet, according to the Intergovernmental Panel on Climate Change. The underlying cause is an abrupt and persistent increase in greenhouse gases (GHGs) that have accumulated in the atmosphere from human activities, primarily the burning of coal, oil, and natural gas. The addition of these gases to the atmosphere is intensifying the natural greenhouse effect, originally identified in the late 19th century by Swedish scientist Svante Arrhenius, which warms the planet by blocking the escape of heat into space, thereby keeping it within Earth's temperature range.

CHAPTER 2

Innovative Technologies for Green Energy

In the last heaved the future of planet cleverness of the planet basic principles for empowering attitudes in such a crucial opportunity the absence of ground as global climate change. The reason is a potential dramatic to dramatically reduce gas emissions costly using green energy, here is strong evidence that global both possible and viable now. Yet many questions remain -- not all climate possibilities concern whether or not the consequences of global warming require fundamental issues of human self-constructed to the power of fundamental knowledge.

Although renewable energy use is increasing rapidly, taking over from new technology to clean energy, global warming, the outdated mind logic for fuel discovery & extraction has. Not only has green energy even bigger potential than we imagined a decade ago, but breakthrough cost-performance illustrates alternative chances for low emission projects. A low-carbon energy boom dramatically reduces the danger level fuel recognition, but the different types of air pollution associated with it, and dramatically reduces the number of global warming. The really going green energy-related drug - both in financial and ecological terms - indicates that exploiting demand-driven technological innovation for renewable energy, non-energy technical solutions to slow world warming can offer equally large dividends.

Solar Power Innovations

Inhabitants of San Francisco who install solar energy systems can offset individual fuel consumption by partnering with PG&E. In 1983, solar energy panels were provided not only by solar collectors at Permanente but also by a requirement of all new construction to fulfill at least 33% of hot water in households. The Solar Energy Research Center has researched ideas for renewable energy sources to lay the groundwork. CSP winners can also be linked to where CSPs are positioned in China, the most polluted Chinese region. Waste and new thermal power towers between 50 kW and 150 kW capacity are also being installed by Southern California Edison and Albright & Bache, Inc. due to proven environmental benefits. Small and medium-sized power plants can deliver two to seven months between commissioning of a natural gas combined-cycle plant and increased need for electric power, with short construction span (12-42 months before commercial scheduling).

Since the advent of solar panels in 1954, solar power (capturing and storing the sun's rays) has become popular in developed countries. One of the key reasons solar power in the US rose 77 percent since 2001 is due to changes in the investment tax credit. At times when the federal subsidy reduced or consumers sensed instability of regulations, the sales of solar systems fell. Worldwide, large-scale solar thermal or photovoltaic plants have already been installed to convert sunlight into electricity at an initial cost of four billion dollars. Popular in Bakersfield, Inland Empire, and surrounding regions, concentrated solar power (CSP) is used in California to supply 354 MW. The land necessary to generate wind energy is greater because of the substantial differences in power density illustrated by the Mojave 100 MW tower with mirrors that focus sunlight. Solar has achieved an exponential increase in the energy impact in Germany (current leader) and Spain.

Wind Energy Advancements

During 1995, modern wind turbines reached a milestone with 3,130 megawatts of electricity generation installed worldwide. The growth of modern wind turbine technology, some installed in the 1970s and still operating today, illustrates the durability of harnessing wind energy. According to experts, cumulative installed wind generating capacity could give wind a 100,000 megawatt perspective in the next twenty years. Competitive with and replacing traditional sources of electricity in many regions, wind-generated power is also clean and efficient. The wind blows most strongly along the world's shorelines, and new wind farms are helping reduce the global reliance on coal.

Wind and water used to be the two main sources of power used by man. The sails on early clipper ships made ships the fastest vessels on the water, but without an adequate power source, they couldn't sail at all. Eventually, steam power and later diesel replaced the sails. Now there's a renaissance of sorts for the power of moving air. Wind farms, concentrating a number of giant windmills in areas with prevailing wind, are turning wind into electricity that powers the coastal and some inland settlements.

Hydroelectric Developments

Currently, the Engineering Energy Costs Rates and Finance Committee has been reviewing Canadian hydroelectric projects' environmental and financial aspects. Within Northern Provinces, the Quebec Hydro (HMN) in large-scale developments includes the

Morley Beck Hydro Development, Chiselab-Pitrecoeuill-Pierre-Blanchet Hydro Development, and the Nicol-Forestville Project. The Hydro-Quebec Register reports that 7000 MW are to be instated by 1989. The Northern Manitoba Hydro Development Program reacted the Kelsey Generating Station, the 100 MW Grand Rapids Project and the construction of the Limestone and Kettle Generating Stations, among others by HMN.

Canada, with its large unspoiled rivers still flowing freely to the sea, is among the few nations that may still harness water for hydroelectric developments. The earliest hydro developments occurred along the Canadian Shield. The Churchill River in Saskatchewan was harnessed in the 1920s. In northern British Columbia, British Columbia Power Corporation's W.A.C. Bennett Dam created Williston Lake, and the Peace River was dammed to supply power into Alberta. Large amounts of power were then diverted to industrial attractants, such as large smelters and pulp and paper mills.

CHAPTER 3

Sustainable Agriculture and Land Use

The Donor Board, in its cutting recommendation CXD 2/12, refers to "decisive" action on food prices. World meat prices rose by only 1% between 1989 and 2012, despite an almost 50% increase in international livestock numbers. The average annual value of meat output per capita in 2010-2012 was $1,129, while world population averaged 7 billion over the period. The 2010 Pentagon-commissioned Rand report published under the title "An Abrupt Climate Change Scenario and its Implications for United States National Security" (and recently made even more up to date by Faye Duchin in an appendix to the book on ecological macroeconomics written in memory of Inge Rngine) rehearses the risk of a global agricultural slowdown as the American breadbasket is hit by investors' withdrawals in response to declining agricultural productivity. Food security depends not just on having enough food, but also on being able to buy it. The simplest policies to increase food security are those which reduce climate change. For those in the rich world who think food came only from supermarkets, changes in diet will be much more effective in tackling climate change than diplomacy and technology.

Humans rely on just 1.7% of our planet's surface for cropping and only 3.3% for raising farm animals, but together these uses are the largest cause of habitat destruction and one of the greatest threats to biodiversity, having already extinguished vaster numbers of species than United Nations forecasts of climate change's impact. Redirecting farm subsidies from maximizing the yield per acre to maximizing the yield per tonne of carbon emissions would achieve a triple win by enhancing food security while restoring ecosystems and cooling the planet.

Agroecology Principles

The work of some developing country research organizations and non-state actors in agroecological practices confirms that agroecological practices can be effective in reducing the emissions from agriculture and transforming the food system – while at the same time increasing resilience to the effects of climate change. The adoption of these technical, policy, and institutional changes will require a combination of solid scientific information, communication, education, and social, political, and economic support at different levels of government. The private sector can and should also play a key role in achieving socio-ecological sustainability through alliances with small-scale farmers and the adoption of more eco-friendly practices in their value chains.

The term agroecology refers to the application of ecological and social concepts and principles to the design, development, and management of sustainable agroecosystems by promoting biodiversity and conserving the relationship between agricultural plants, animals, humans, and the environment in the face of modern industrial technologies. Today, the success of agroecological practices, including integrated pest management and biological control, or agroforestry systems that incorporate crops and trees, is being challenged.

Permaculture Practices

Permaculture practices offer benefits in local, social, and global contexts, whereas Kyoto mechanisms once again turn complex issues into an economic market model that offers few benefits in any of these three contexts. Sustainable management and co-benefits thus form key elements within the broader framework of permaculture theory. The principles and practices of permaculture have the potential to sequester atmospheric carbon within a multitude of land use systems, grassroots agencies, human settlements, and economies at local, regional, and global scales. These benefits accrue while providing co-benefits addressing a range of environmental concerns, by improving the effectiveness of current environmental treaties and mechanisms of the approaching post-2012 first commitment period of the UNFCCC.

Permaculture is a holistic design system concerned with producing a sustainable food, water, shelter, energy, and community function. It does this by integrating natural systems, waste management/ prevention, energy efficiency, biomimicry, community planning, and a development cycle that aids in the creation of regenerative systems. Permaculture systems are modeled after Indigenous land management practices rather than techniques such as small-scale organic gardening. Ethical decision making is a fundamental element of permaculture. A practitioner understands that they are not separate from any system or resource base, but are very much a part of the whole. They then look to design for long-term beneficial relationships amongst the components of the system, such as between humans and other living organisms, the flow of energy and resources among these components, and the regulation of these processes, and the alignment of these goals with the rest of the natural world.

CHAPTER 4

Green Transportation Solutions

The Climate Solution: Innovative Approaches to Combat Global Warming; by David Hodgson; published by The Foundation for Global Community.

Green Transportation Solutions

The Problem: Transportation (including shipping) accounted for a large part of the 6,000 million tons of CO2-equivalent greenhouse gases emitted in 1997. High rates of miles-per-gallon in portions of the developed world hide the greater gas consumption of the ever-increasing number of less-efficient and carbon-intensive automobiles and trucks packing the streets and highways of most of the planet. There are several challenges in reducing carbon emissions from transportation, including the very large number of gasoline-consuming vehicles in the world. Americans buy nearly twenty million vehicles per year, and the overwhelming number of miles driven are in cars and light trucks, particularly the latter.

The Solutions: Solutions such as improving gas mileage seem an obvious approach, but past experience shows that as automobile engine efficiency improves, Americans simply decide to buy larger and more powerful cars and trucks. What is required is more radical thinking. For example, we should develop urban settings that encourage walking and are friendly to bicycles and efficient public transportation, which has many benefits over the car but has been demonstrated to be inconvenient when cities are designed around cars. And as we design more automobile-oriented environments, we should ask whether it is really necessary to back up every single car in the United States with a high-tariff oil economy that forces us into an unreasonably high level of military activity, allows petro-dictators to provoke the industrialized countries, and destabilizes major portions of the Earth's population. Car sharing in areas near convenient public transportation will lead to still more efficient use of energy consumed in transportation. The quality of air in our urban environments and the public health benefits of having communities built around people's needs rather than automobiles' demands would make such communities even more attractive.

Electric Vehicles and Infrastructure

The accelerated introduction of EVs offers a number of potential benefits, including significant greenhouse gas (GHG) reductions, improved oil security, and increased use of intermittent renewable resources. By 2050, the U.S. has the potential to reduce LDV GHG emissions by 75% through electrification, assuming technological, economic, and policy hurdles are dealt with in a timely manner. The potential reductions in use of oil, which are heavily fueled through imports, are also significant; with the continuation of historical trends, the U.S. is projected to continue increasing its oil consumption. Increased vehicle electrification could help bring the U.S. back toward reduced oil use and associated geopolitical vulnerability.

Electric vehicles (EVs) offer an enormous opportunity to reduce transportation GHG emissions. Long-term, EVs will be charged with electricity increasingly generated from low-carbon resources.

However, in the short term, significant emissions could occur due to increased electricity demand from a coal-intensive grid. The pathway to EVs yielding significant GHG reductions is clearly tied to changes in the electricity sector. We can explore opportunities for additional near-term GHG reductions through efficiency measures that reduce overall energy consumption as well as grid interactive transportation technologies that can capitalize on more low-carbon fueled electricity.

Public Transportation Upgrades

Improved regional and intercity passenger rail results in even more dramatic reductions in greenhouse gases. One mile of commuter rail can transport over 600 times the number of passengers as the typical single-occupant vehicle. The 134,000 miles of passenger rail in the U.S. carry the same number of passengers as the Empire State Building, but sixteen times the size of the building would be required to park the single-occupant vehicles that are displaced by passenger rail. With broad and deep routing, rail can serve multiple metro areas, provide reliable service to downtown areas served by mass transit, and reduce passenger transportation GHG emissions.

Bus rapid transit (BRT) systems are affordable and easy to deploy, but the payoffs are immense. A study of 21 cities throughout the world found CO2 reductions of 17 percent in South America and 35 percent in Asia only seven years after the introduction of BRT in some of the cities. BRT is a "little brother" to larger, heavier, and more expensive rail systems. Smaller cargo and people movers circulate on relatively narrow tracks or roads with protective restraints such as curbs, bollards, fences or other separation devices. Investment costs are 1/10th that of a light rail system or 1/100th of a subway, and a 20-mile BRT line can be operational in two years at a cost of a few hundred million dollars.

CHAPTER 5

Climate Policy and International Cooperation

The Kyoto Protocol already brings together financial mechanisms designed to transfer resources to developing countries to implement a range of projects, mostly related to improving energy efficiency and removing deforestation. Renewable energy is expected to play a crucial role in the global electricity system in the coming decades, not only because they are essential in combating climate change but also because they are more economical resources than conventional hydrocarbons. The technological development in this sector has been very significant in recent years, with a significant fall in investment costs and a substantial increase in efficiency. This is reflected in the increasing weight of renewable energies in global energy production, yet it does not lead to the reduction in global carbon emissions needed in the coming decades.

International cooperation is essential for regulating global emissions of greenhouse gases and ensuring compliance with the agreements reached. In the European Union, for example, various greenhouse gas emissions regulations, such as the use of more energy-efficient durables and the energy decarbonization index, were discussed to ensure plummeting carbon dioxide emissions while also maintaining its role as an international leader in combating climate change. International cooperation not only includes the negotiation of new and effective treaties reducing emissions and regulating emissions trading but also jointly developing technologies to reduce emissions. These technologies may create many opportunities for developed and developing countries.

Paris Agreement and Its Implications

The latest global meeting to combat climate change was in Paris in December 2015. Nearly 200 countries around the world came together at that time and agreed that the world needs to cap the rise in global temperature at not more than two degrees from what it was before the industrial revolution in the 19th century. The Paris Agreement is a predecessor to another global meeting, the Copenhagen Summit, which took place in 2009, where 193 countries agreed to the goal of limiting the rise in global temperature to two degrees, and ideally to 1.5 degrees. This time, however, 187 countries agreed to limit it to one and a half degrees, and six countries refrained from doing so. Putting an upper limit of one and a half degrees was a big win from a developing country point of view, especially when big contradictions persist between developing and developed countries with regard to mitigating against climate change.

Global Climate Summits

Given the importance of timely and appropriate regulatory action, real progress in climate change resource management cannot be made exclusively at the international level. Yet each annual leadership moment is frequently missed because of an overabundance of concern raised about the high cost of dealing with climate change. Participants have a deep interest in ensuring that a deal is struck that maximizes chances for bipartisan U.S. Senate ratification. In terms of balancing the climate deal with costs, it is not clear where the maximal possible politically palatable ambition is - the chances that U.S. legislative politics will automatically translate into and be limited to the politically possible, it is one thing that the participating nations share independently of the content of the emerging agreement.

Whether or not a global climate agreement is ultimately achieved in Copenhagen, the decisive moment for global action in the fight against climate change is the annual United Nations climate summit. At these yearly gatherings, the nations of the world chart their course toward a post-Kyoto Protocol climate regime and agree on nearer-term actions to be taken. Leaders from around the globe, including heads of state, cabinet officers, and other high-ranking officials, spend intense days or weeks negotiating to advance national and international climate policy efforts. Over the years, it has become accepted wisdom that the highest risk time for progress at the climate summits is the tiny window of time during each that the head of state - usually a president or prime minister - is around. At that moment, the world experiences the most intense interest and attention to the international climate change issue. That attention is critical to people's understanding of the problems associated with dealing with climate change and, ultimately, to finding an actionable solution.

CHAPTER 6

Community Engagement and Behavioral Change

Community action is one effective way to build support for government action to reduce greenhouse gas emissions, as well as to achieve emissions reductions on a smaller scale. Research on community action to mitigate climate change provides some guidance for designing strategies that will be most effective. Communication tactics that focus on achieving increased awareness of climate change, encouraging concrete action, and changing attitudes are most effective. Information alone is not enough. Strategies that provide nonfinancial incentives for participants, such as increased social approval or environmental quality, can be more effective than providing financial incentives. This suggests that voluntary programs, which have the advantage of achieving a greater degree of participant engagement, may be more effective than regulatory programs for many activities. Finally, personal communication from community members and leaders, and providing opportunities for collaboration and information sharing among community members, are particularly effective strategies.

Public Awareness Campaigns

As a result, the climate change debate with opposition to climate science that is both unnatural and morphs the issue from a shared problem into a pitched battle that just has to have a winner and a loser. People can be either for the environment or for our economy, but not both. Because of this, it is important to be aware that calling people on their environmental self-interest frames the issue in such a way as to make enemies of what should be allies. There is work to do in moving conversations away from stigmatized discussions and winner-take-all arguments. Alongside working politically on this issue, it is also helpful to create educational public awareness programs and make them available to anyone.

Will Bruyen, a marketing and sustainability expert, created a social marketing strategy to help people become less defensive and start thinking for themselves so they can make intelligent decisions about the climate crisis. According to Bruyen, talking about the climate crisis is politically loaded and involves stigmatizing. People see environmentalists as radical, self-righteous, anti-business, anti-jobs, and anti-civilization. In the past, environmentalists were radical, self-righteous, anti-business, anti-jobs, and anti-civilization, but that was when talking about the environment consisted of boring and scolding over message and threat-based communication techniques.

Community-Based Initiatives

Reducing emissions may also take the form of voluntary actions by community organizations and citizens. Currently, a range of community-based initiatives already contribute to reducing greenhouse gases either directly, through the projects that are undertaken, or indirectly, through the potential of reinforcing a community culture that is concerned with environmental matters. The methodology and research project on Communities, development alternatives in the new millennium by World Resources Institute recognizes that local initiatives can potentially play a significant role in global environment management. Many organizations are already following this route and people worldwide are beginning to take practical steps to halt pollution at the source. They recycle, use energy-efficient appliances and transport models, and invent safe substitutes for hazardous chemicals in factories and products used every day. These initiatives are mainly centered on the actions that communities may take as consumers. The many efforts that are valuable but that focus on more complex activities are not included in these statistics. The UNCED process needs to endorse fully the partnership between central governments and other authorities – such as local government, industries, multinational manufacturers, non-governmental organizations, local communities, and citizens – which is so critical to achieve environmentally sustainable development, as recognized during the opening plenary session of the North American regional conference 'Facing environmental challenges' and stated in the Ministerial Declaration of the 17th session of the UNEP Governing Council meeting, where the pledge to activate this partnership is reinforced.

CHAPTER 7

Corporate Responsibility and Sustainable Business

A growing number of the world's largest companies have made substantial commitments to energy efficiency and renewable energy - the twin pillars of a climate-friendly energy future. Companies that design, build, and use efficient and clean energy products and services are striving for what amounts to insurance: they are mitigating environmental risk and reducing uncertainty. In the process, thousands of households, manufacturers, commercial buildings, and utilities are better for it - creating jobs and boosting local economies. These corporate sustainability leaders are enhancing their bottom lines. Now, the urgency of the climate crisis calls for a broad-based response from corporate America and from green businesses worldwide. Industries stand to win by steering the world economy to a low-carbon, more climate-friendly path. With a hodgepodge of national regulations likely to emerge, industry stakeholders can encourage the federal government to seize the opportunity to negotiate a comprehensive climate protection agreement with other leading nations.

While government policy offers the greatest opportunity to greatly reduce global warming pollution, many corporations have recognized the climate challenge and have developed profitable business practices and operations that actually slow, rather than aggravate, climate change. In the United States, almost 500 companies support the federal government's standard-setting efforts to build better cars and trucks. Over 425 companies have urged action by Congress to set mandatory limits on global warming pollution. More than 180 companies have written a comprehensive plan to cap and reduce global warming pollution cost-effectively, using a market-based approach to tear down artificial regulatory barriers and speed the transition to environmentally-friendly, cleaner energy products and processes.

Eco-Friendly Supply Chains

The concept here is simple: To maximize energy consumption and carbon emission reductions, supply-chain opportunities for energy efficiency must be addressed in the same way as traditional issues—cost and service—to produce a robust, balanced solution. The following presents three examples discussed by supply-chain leaders that illustrate the potential of these methods: transporting goods over long distances using the most energy serious modes available, such as trucks and planes; packing goods for shipping in oversized containers, which reduces congestion and lowers peak demand and energy consumption.

Some transportation and logistics companies offer "green" supply chain programs to help their customers become more energy efficient and, as a result, cut their carbon footprint more deeply. Management consulting firm A.T. Kearney, for example, has an outsourcing joint venture called Xelnetwork that offers a line of eco-efficient logistics services, including consulting, helping customers cut energy consumption and carbon emissions. These services may include facilities layout and design within the transportation infrastructure, warehouse layout and design, process design, supply chain planning system, and performance management. Furthermore, aplic also provides carbon transport calculators, providing data for environmentally friendly supply chain decision making.

Carbon Offsetting Strategies

Assembly Joint Resolution 86, passed by the Nevada State Legislature in 2009, supported and endorsed voluntary carbon offset protocols as an effective means of reducing greenhouse gas emissions. Agencies and institutions in the State of Nevada have taken a lead in offsetting their respective carbon emissions. However, most heat-trapping greenhouse gases are not a local but a global concern, provision of global benefits, financial burden, and thus, global benefits. Voluntary climate service activities help to establish the commons that are required for the global implementation of technologies and management techniques that may be developed to combat heat-trapping gas emissions. It reflects private partners' willingness to pay for avoiding any further damage and support GHG record removal, protection of capture, and international cooperation.

Offsets are reductions in emissions of greenhouse gases made in one place to compensate for the emissions taking place elsewhere. Many offsets represent verified greenhouse gas reductions from an adoptable action. Carbon offsetting is generally achieved through financial participation and improvements made elsewhere, that is, one finances what is not done directly, but which is something that produces a benefit representing a global climate service and the respective reduction.

CHAPTER 8

Climate Resilience and Adaptation Strategies

At the foundation of the current heightened interest lies the awareness of a growing number of villages, towns, and cities near and far that interact with their ecological resources for daily existence and long-term sustainability of water use and ecosystem resilience. Tools are available today that can enable communities, often in collaboration with each other, to ensure that their water systems are optimized. These tools, such as LIDGuidance, provide the information and expertise that enable authorities to develop the potential for enhancing treatment and infiltration of water. At the same time, climate change leads to shifts in the distribution of precipitation and snowmelt that can change the magnitude and duration of streamflow, which may change the relative impacts of stormwater discharges.

The role of the built environment in ecological stability and public health continues to evolve, and minimal interactions have been incorporated into infrastructure planning. However, engineers, planners, ecologists, and policymakers are beginning to realize the multiple benefits in designing the built environment to interplay with ecological systems and processes that will handle the major impacts of human habitation. The ability of nature to cleanse wastewater, sequester carbon, buffer storm surges, and reduce flood severity, etc., had remained understudied but is increasingly being studied. Communities across the USA and the world are increasingly in need of these natural services. While the potential has long been known, it is only recently that operational integration of such services has been investigated and applied.

Nature-Based Solutions

Trees are vital to the health of the planet as they absorb carbon dioxide and emit oxygen. Trees remove up to 30% of the carbon dioxide from the air when they are mature, and while woodland and forest cover is increasing in some areas, deforestation is still a problem. Research shows that pairing trees with standard shallow-tillage agricultural practices can boost the amount of carbon absorbed from the atmosphere. Coastal and marine ecosystems also play a key role in carbon sequestration, but the loss of global seagrass meadows combined with high greenhouse-gas emissions is leading to reductions in their important ability. Economic strategies for mangrove and wetland restoration can restore marine ecosystems and increase carbon sequestration.

There is great potential for use of natural systems in a sustainable, cost-effective way to absorb carbon and stave off climate change. Nobel Laureate Al Gore has raised awareness about the importance of 'drawdown'—reducing greenhouse gases in the atmosphere. The concept has been taken further by Project Drawdown, a collaborative effort of leading researchers around the world who utilized systems thinking and modelling to analyze and map the 100 most substantive solutions to address global warming. The ranking of solutions includes reducing food waste, building with wood, and protecting forests.

Infrastructure Resilience Planning

The utility regulators are precise on the use of ratepayer funds, but no attention is currently placed on the high cost of resilience repair after disaster damage is incurred. The Community and Regulatory Response Model for the prediction of post-sustained and disaster extreme event grid failures was applied to an upgrade decision making by California's largest grid owner, Pacific Gas and Electric when deciding on installing overhead and underground transmission between two substations. By analyzing whether PG&E and stakeholders made the correct decision to underground its system, the gains from developing these advanced methods in guidance for critical infrastructure resilience planning toward desired goals can be demonstrated.

The most critical yet aging infrastructure in the United States continues largely to be ignored in all adaptation planning—our nation's energy, transportation, and water systems. Utilities are required to handle predictable sustained and extreme natural events, but the onset of these expectations has been rapidly evolving due to the frequency and intensity of extreme weather events in response to climate change. The utilities, communication companies, and owners/operators of the rights-of-way connections that connect grid, aqueduct, and transportation networks rarely devise large-scale resilience planning, which is typically the responsibility of a diverse range of government organizations. The interdependency of networks and the fact that they were built in different times and under different regulatory constraints heighten difficulties in developing these plans.

CHAPTER 9

Education and Training for Climate Action

The specific needs of elementary, secondary, and tertiary students vary, just as they do for people already in the workforce. Primary and secondary students can prepare for an uncertain future by learning the skills needed to acquire, critically appraise, and use new knowledge. University-level students should also understand the need for system-wide change, laws and regulations, human behavior, and political change—adult students must understand the importance not only of policy change, but also of stabilization of atmospheric GHG concentrations. Management and communication are also important across the board. The ultimate goal is to empower individuals around the world to shape a low-carbon future. This will help prepare and educate a broad, diverse cohort of present and future leaders in the fields that are critical for implementing climate solutions.

Climate change is much more than just an environment or energy issue. It has implications for public health leadership, human rights and poverty alleviation, equity and justice, disaster resilience, and other areas. It alters the very future of the planet. As such, it requires meeting the educational needs of a diverse group of students—not only high school science classes, but also undergraduate and graduate programs, and the needs of adult learners. The goal is for all students to understand climate change impacts and solutions. As regards people already in the workforce, training and re-training will be needed at different scales. Many see undergraduate and graduate-level programs as essential elements of broader climate change policies. Education can help reduce likely future conflicts by increasing the number of technically capable individuals able to lead the societal transition.

Climate Literacy Programs

"The SMART School video network creates greater global awareness and ensures every student's voice by using the visual language of computers to better see, better hear, and better understand the actions of humankind to live together sustainably on the only habitable planet, Earth."

There is a great need to increase the public's understanding about managing the risk associated with global climate change. We need to provide better resources for students, educators, the media, and for all citizens about how to think constructively about the important issues surrounding global climate change. The public is often excluded from many policy decisions, thus limiting their collective support for a policy's implementation. The National Research Council calls for states, corporations, and environmental organizations to increase their efforts to raise the public's understanding about global climate change and advocate for "prudent and ethical policies." Great care and guidance is needed when developing climate education materials.

Professional Development in Sustainability

Many educators left the program with an evolved sense of the interconnectedness of the people and the natural world. Environmental educators often come to their jobs out of a love for nature and an intention to protect it—a perspective that enjoys a relatively high degree of intellectual self-actualization. We broadened participants' sense of self and responsibility to the community to include people as well as the rest of the biophysical world and provided them with the competencies they want to have to be more activist-oriented. Many educators were re-empowered in the personal sense of contribution that culminated from their experience working with SENCER. They can see many connections about the biophysical world and reported an increased intellectual eagerness to learn new complex scientific knowledge about real-world questions. Many educators reported gaining a greater appreciation of and deeper willingness to do interdisciplinary work in the area of sustainability and that they were interested in interdisciplinary place-based learning. They took back a new appreciation for the community and the many benefits that came from their engagement. This included a revised personal professional role from that of teaching science as counting facts to helping participants out with skills and dispositions that could lead to a better quality of life in the midst of very difficult civilizational and world choice. This is evident in their teaching preparation classes but not in the deep and real-world long-term engagement in which these thinkers need to do.

We conducted a needs assessment at the semiannual gathering of environmental educators at the center. We discovered that they shared a need for more advanced knowledge, skills, and dispositions for teaching others about sustainability. This training provided them with powerful new tools for engaging diverse learners, imagining innovative career paths for themselves, and making ongoing contributions to their professional practice and communities. The framework of the program is built upon a comprehensive treatment of the issues of sustainability and a comprehensive module on learning and teaching strategies. We provided participants with an in-depth understanding of larger systems of life that create community sustainability, including a deep understanding of the interconnectedness of people and the biophysical world in which they are a part. Communities are central to the understanding that groups of individuals of multiple species form and that they have been and continue to be dependent upon other forms of life.

CHAPTER 10

The Role of Indigenous Knowledge in Climate Soluti

But the largest part of the world is yet to recognize it. For a country like India, which will depend on 65% of its food crop on the winter monsoon, the importance of the link between the climate and traditional wisdom used to design management methods is amplified. If a "hey policy makers in the world" ear were to come close enough to the ground, there was substantial evidence of how crop varieties that are sown when the monsoon of the Indian Peninsula is good end up providing sustainable agriculture despite the highs and lows in subsequent years. Monsoons all over the peninsula are a meteorological certificate – a standard index that will not change while the ENSO or the Indian Ocean Dipole or any other extended-range forecast changes on a weekly to seasonal scale. In other words, the climate is 'inside out.' And those 'tomorrow' of monsoon patterns know that months in advance of whatever enso baskets. Such knowledge from key dimensions could suggest the patterns of changes in the distribution of biodiversity.

In India, for example, the NCF has realized the practical wisdom of rural communities for the conservation of traditional crop varieties when adverse weather disrupts minor concerns. This program provides initial support to bring communities together in the form of providing knowledge-sharing platforms to enable them to make decisions about whether the crops are at risk. This ancient wisdom is super modern, and the results of such initiatives are supported by the climate modeling and data modeling world.

The world's indigenous people, whose wisdom and knowledge span millennia, have, over the generations, adapted farming and hunting patterns to retain the rhythm of life. This deep well of knowledge offers so many lessons on how to maintain ecosystems, species, water, soil, and climate that the world can use in our efforts to oppose climate change.

Traditional Ecological Knowledge

Atmospheric and climate data have major use to many subsistence and traditional rural communities, and it is increasingly recognized that these communities also have something to teach. Both share a common ingredient: precision observation and reporting of the natural environment. Outlooks are usually expressed in terms of patterns rather than concrete numbers, mirroring the limits of the scientific process and the available data. This should be particularly valuable when evaluating climate variability or near-term weather prospects in remoter areas. It has implications for emergency management and extends beyond agriculture to other activities, such as fishery and hunting management, forest and water resource management, public health, and adaptation practices.

Partnerships for solutions can reflect the integration of traditional ecological knowledge (TEK) and Western science, with communities stressing ecological balance as they adapt to new realities. Many indigenous peoples' experience of tradition often involves change. In traditional practice, there is often little caution against experimenting with such change, as long as it is undertaken within the existing cultural framework and within a spiritually oriented moral and ethical framework. Experimentation may have been more noticeable in a world before the centralization of power, when small groups could learn rapidly by empirical methods and share experience beneficially. Much can be learned from the community-based experience.

Community-Based Adaptation Practices

Throughout the world, many of the most motivated groups get down to business with a minimum of resources. Internationally, a growing number of alliances, networks, and other forms of coordination are linking a large and diverse group of organizations and communities dedicated to promoting such innovative approaches to local sustainability. Collectively, these groups do not talk of mere adaptation to a warming planet but rather embark on development and, in some cases, full-scale community organizational change that leads to a near-term reduction of atmospheric greenhouse gas concentrations.

Many governments and corporations are not enthusiastically tackling global warming. The strategies employed by the big players tend to be high cost and more about financial show than carbon emission mitigation. Still, there are good people and innovative street-level solutions throughout the world. In many cases, these are some of the up and coming successful large-scale solutions. With a bit of tending, these grassroots solutions could grow and become a critical balance to bloated, heavyweight government and corporate environmental strategies. Numerous community-based responses to climate change demonstrate the willingness and creativity of local people to evolve new ways to cope with their altered environment.

CHAPTER 11

Innovative Financing Models for Climate Projects

Another area we touched upon was reducing leakages of these funds, to help sustain the investment. In many parts of the world (especially the low-income countries) corruption is a primary institutional risk for climate finance. In 'developed countries' the systematic loss of precious money in tax havens is a significant wealth of complication. The press has revealed several shocking thefts in Nigerian oil businesses.

While many discussions on innovative financing mechanisms for climate change initiatives often focus on funding new technologies or projects, our workshop also discussed topics that may be considered more boring and long-term in nature. One of the mechanisms touched upon quite often during our discussions was raising the monetary value of carbon, a necessity towards reducing greenhouse gas emissions. The mechanism of Funded Carbon Credits (FCCs) was proposed, which when incorporated in the Clean Development Mechanism (CDM) could synergize actions by nation states, the private sector and the UN, which in turn facilitate the financing of adaptation and avoid the risk of non-invention of future technologies.

Green Bonds and Impact Investing

If you are new to impact investing, "A Short Guide to Impact Investing" is worth reading. For an overview of the proposed environmental benefits framework for the Green Bond Principles, it shows the range of potential projects in which the proceeds of green bonds can be invested. Many of the listed projects have considerable climate co-benefits, including energy efficiency, sustainable water, wastewater treatment, public transportation access, carbon reduction goal alignment, renewable energy, and sustainable land use. Forest resiliency projects connected to wildfire risk management create an important co-benefit of enhanced water security. In other words, virtually all the projects in the proposed environmental benefits framework also advance some part of the global warming solution.

Green bonds and impact investing, a relatively new development, have the potential to drive billions of dollars in capital to climate mitigation and adaptation projects and to help individual investors "do good while doing well." With American taxpayers facing trillions of dollars in bills because of extreme weather events and other climate-related costs, these new investment vehicles could not be timelier. Green bonds are similar to conventional bonds, with the key difference being the use of the proceeds: Green bonds must finance projects that provide clearly demonstrable environmental benefits. Impact investments are a variety of capital market tools that offer for-profit results to investors but also provide funds to employees and communities. Results are typically measured in terms of social and environmental metrics as well as financial return. They are a relatively new addition to the climate finance toolkit, but in the United States and globally they are growing rapidly in importance.

Carbon Pricing Mechanisms

Recognizing the importance and urgency of a novel carbon pricing instrument that can truly help reduce carbon emissions and lessen economic impact, this book proposes such mechanisms. This includes a perpetual carbon exemption life annuity (PCELA), a preferred emissions credit, and a carbon tariff à la border carbon adjustment that invests in developing nations and effectively reduces global carbon emissions.

The international pressure to curb greenhouse gas emissions has grown, forcing nations who had already implemented carbon pricing instruments to raise their established marginal carbon pricing levels. But while carbon pricing can steer behaviors towards sustainable and low-carbon alternatives, it can generate significant economic costs for the public if the concept is inequitable.

The International Monetary Fund and the World Bank Organization reported that the number of these instruments tripled in the previous decade, with significant momentum in aligning carbon pricing with the Paris Agreement. The carbon reduction levels have made headway, and the Net-Zero Carbon Coalition also recognized carbon pricing as "essential to decarbonize the global economy."

Carbon pricing is an innovative approach that places a monetary value on the world's largest source of greenhouse gas emissions: carbon. In 2021, more than 64 carbon pricing instruments were in place around the world, covering 22 percent of global greenhouse gas emissions. These market-based policies either charge emitters for the carbon they produce or reward them for avoiding carbon emissions. Over time, carbon pricing helps companies innovate, reducing their emissions and that of their products, services, and technologies.

CHAPTER 12

Technological Innovations in Carbon Capture and St

Capturing carbon and storing it indefinitely in the earth seems far-fetched. Can it happen? On the capture side, good progress has been made, and large-scale demonstration projects are planned. Major barriers but few technical showstoppers remain. At some foreseeable cost, we can capture almost all the carbon dioxide a society uses or creates, removing most of the climate threat to a fossil-powered future. Numerous promising capture concepts and companies are attempting to create complete capture of carbon from power stations, and they appear to have a good chance of overcoming the capture challenge.

Carbon capture, as the first step in CCS, is an immense technical challenge, involving vast flows of air or oxidizing gases through immense apparatus to extract minute quantities of carbon dioxide. CCS is the even greater challenge of vastly enlarging this infrastructure to roughly the same scale as the present worldwide oil-and-gas industrial systems, and storing away from the atmosphere some 29 times the volume of fluids now extracted and processed, in stable, concentrated form, possibly for thousands of years. The first commercial capture process began operations in September 2014 at the Canadian Boundary Dam Power Station, owned by SaskPower. This carbon capture process will capture and store a maximum of 1.3 million tons of carbon dioxide per year from an existing coal-fired power plant facility.

Direct Air Capture Technologies

Research on DAC has intensified significantly over the past few years because of the growing recognition that even if we manage to execute an ambitious near-term mitigation strategy, it will be necessary to engage in substantial negative CO2-emissions efforts sometime after 2050 to maintain a climate with acceptable levels of risk beyond the year 2100. Even though there have already been several demonstration plants in Europe, Canada, and North America that can successfully remove CO2 from ambient air, the limitations on practical scalability, energy requirements, and costs of these technologies are proving difficult to overcome. However, it is worth noting that certain industrial processes which could implement low-cost CO2-rate refinement will cover about 10% of the global CO2 emissions while DAC would provide a guaranteed negative CO2 emissions pathway.

Direct air capture (DAC) refers to a variety of technologies capable of removing carbon dioxide (CO2) from ambient air. Over the last decade, multiple start-up companies have developed ways to physically bind or chemically absorb CO2 molecules from diluted suspended particles. For several years, Jan Wurzbacher and Christoph Gebald at ETH Zurich in Switzerland have run a research group focused on liquid sorbents investigated for energy-efficient, large-scale DAC. Central to their approach was humidity swings and temperature swings. By altering the internal chemistry of a porous, inert solid with changes in either humidity or temperature, CO_2 gas uptake occurs, allowing CO_2 extraction from the ambient air.

Enhanced Weathering Processes

In areas where the production of basalt by cooling of the magma is estimated to occur at a long time in such a way that CO_2 can be safely sequestered, these massive natural flows of basalts are the cheapest existing reservoirs for vast quantities of stabilized CO_2. The practical and economic are many, so that a rapidly increasing number of mitigation approaches can be put into the fields.

The technology for achieving highly energy-efficient processes to mine massive quantities of alkaline earth silicates, located in the whole world for use by injecting CO_2, sequestering CO_2, expanding, and then returning the products of CO_2 neutralization with a closed-loop process. The same basic processes can be used in a distributed way if heating is required to initiate supercritical reaction of the rocks with CO_2. Due to many environmental factors, specialized missions will be favoring the commercial development of ample CO_2 sequestration opportunities.

One way is to remove and contain CO_2 from coal power plants in regions where substantial environmental benefits might accrue. Ocean mining of atmospheric CO_2 by transoceanic ships, which with numerous pumps deliver very low partial pressure CO_2 to the ocean surfaces, while exploiting differences in solubility of CO_2 with latitude and longitude. The approach for minimizing the potential ocean acidification involves blending a mildly alkaline hydrochloride solution with the sequestered CO_2, enabling the production of limestone.

Another way to enhance the natural chemical weathering process is to use emerging and not-so-new techniques to increase locally and even regionally these processes. By doing this, we can provide multiple benefits to local people, enhance natural systems, and buy time to stabilize the earth systems through the development of alternative clean energy generation processes and energy-efficient strategies.

CHAPTER 13

Urban Planning and Sustainable Cities

Through localized market activity, cities can shape industrial activity to create sustained development and ensure it is achievable without creating high emissions energy production. These efforts require time to assess feasibility, time to build the infrastructure, and time to train a productive workforce. To support these activities, planners need the autonomy to use their knowledge of what is needed for their area of concern. The requisite autonomy must be given by state and national governments for prudent economic development. The required autonomy to plan and execute must be extended to city governments, particularly in places where national policy is extremely restricted.

Cities are an important part of the strategy to address climate change since they contain over half the world's population and are responsible for over 70% of global CO2 emissions. The impacts of climate change, such as impacts on food and water supplies and also heat-related injuries, are felt more acutely in cities than in other areas. In fact, densely developed cities can have a lower energy per unit area, and strong urban systems can result in reduced per capita CO2 emissions. Urban planners must consider planning of residential, energy, and transportation infrastructure and apply land use policies to optimize future societal considerations and energy consumption.

Smart City Concepts

Smart cities must exchange resources with the rural environments that surround them. The principle for slowing down climate change is that the agricultural environment should bind carbon, while the city environment should sequester it. However, well-known political, economic, and legal limitations complicate the establishment of constructive relationships between these exchange quantities. The first critical aspect to understand and manage is related to the carbon balance. City soil is a huge carbon sink. The metabolism of a city is very important for carbon sequestration, storage, and release. Urban issues, like policy and water, can affect the way city soil may release or store the carbon.

Smart city concepts consider the city's influence on climate change and its potential to slow down global warming. The increasing concentration of people has the potential for new relationships between them but also generates undesired environmental effects. It is increasingly obvious that large cities and metropolitan areas offer particular civil and environmental advantages, but they also generate a disproportionate impact on the environment. Besides having 55% of the population, they produce 75% of the world's energy use, 79% of the carbon dioxide emissions, 62% of the residential solid waste, 58% of municipal water use, and 78% of the world's CO2 production.

Green Infrastructure Designs

Also, the cycle of unsustainable requirements led to numerous non-renewable building problems. The green infrastructure of biophilic trees with small crowns and numerous hydraulic branches can reduce UHI and impervious surface (capable of absorbing and releasing up to 45% of incoming solar radiation), and can improve air quality, reduce Kaohsiung Municipality ambient temperature, and provide convenient and comfortable atmospheres. In that case, the rethinking of the public space systems from an interdisciplinary perspective, focusing on the critical nodes of physical, human-defined biological, chemical, historical, and cultural positions in the urban public space, has revealed that these gaps can lead to many convergences and complement each other.

Green infrastructure possibilities, especially in the design of "biophilic" roads and parks, and discusses an integrated dynamic-couple IoT- and 5G-enabled.

Urban heat islands (UHIs) are due to significant impervious surfaces such as cement or asphalt parking lots and roadways, the absence of vegetation, and the domination of buildings. Traditional cityscapes are covered by nonreflective hard surfaces, heat-trapping buildings, and inorganic infrastructure, which retain warmth during the night and prevent the cooling process. Ocean breezes are drawn into city streets, further reducing the city's overall ability to cool down, as vehicle traffic damages the evergreen tree windbreak in the urban public space and reduces the remaining number of trees shading city streets. UHI can have less direct and far-reaching consequences in other realms. For example, enhanced air pollution and greenhouse gas emissions, increased air conditioning, augmented heat headaches, construction workers' heat stroke, water quality decline, power failure, or building failures.

CHAPTER 14

Ecosystem Restoration and Biodiversity Conservatio

Focused investment and new laws can protect the planet's most important remaining natural forests. With protection, these powerful catalysts of climate stability will delight future generations, reduce current impacts of moisture lost from climate instability, and provide us with irreplaceable ecosystem services forever.

The atmosphere's capacity to support human economies is enhanced by natural forests and destroyed by human economic activity. Conservationists cannot effectively talk humanity out of anything that can be bought with $9 in a world running a deficit on a one-time supply of $6 to $53 trillion. Furthermore, forest protection is a hard sell in a world where hundreds of millions of people suffer hunger, malnutrition, and disease, where perhaps as many as a billion people do not have clean drinking water, and where environmental destruction is already cost-effective in increasing the wealth of many of the more than 6 billion global poor.

Protected natural areas are consistently among the top tourist attractions on every continent. They are beautiful places. Yet relatively little land and ocean is protected. It is irrational for humanity to target a gradually increasing fraction of wealth for the aesthetic business that results from only a few thousand dollars of annual solar income. In addition, the planet is now spending its natural inheritance with the understanding that the accounts are holding a different kind of asset - ecosystem services.

Ecological systems attract a growing public following. The public is drawn to solar-powered houses, hybrid cars, wind power, organic gardening, and conservation lands. However, in keeping with business-as-usual, humanity annually turns into consumption and waste 40% of the new energy captured by coal, oil, and natural gas. It protects only a fraction of the biologically productive land and oceans, and it supports reduced regulation rather than protection of species.

Reforestation Initiatives

There are several issues that must be considered when making species selection. The best selection for the species to use will depend on several factors, including temperature, availability of water and nutrients, diseases, and types of pests in the region where reforesting is to take place. A small number of tree species may need to be chosen to cover a broad range of environmental conditions and for the expected long-term survival of the planting. The methods that are generally used to plant new forests are based on traditional methods for sole crop or grain planting. This is effective in terms of increasing the health of the planting. However, because of fire, droughts, deaths, and the general limitations of soil nutrition, planting will be far from optimal.

Reforestation is a time-honored and proven method for removing CO2 from the atmosphere. Providing a depleted soil and the newly planted trees do not release too much carbon, newly planted forests grow rapidly and will remove about 200 metric tons (t) of CO2 from the atmosphere by the time they are mature, about 100 years after planting. While reforestation initiatives have the potential to make a large impact on the CO2 concentration of the atmosphere, the maximum rate of removal is limited by the rate at which trees can grow. The most effective way to maximize the CO2 removed using this approach is to avoid deforestation, allow surviving forests to recover, and, secondarily, to plant large areas in appropriate parts of the world, concentrating on a moderate number of carefully chosen species that have a high growth rate and a long lifespan. The most effective way to remove carbon from newly established forests is to use appropriate species of trees, under the best conditions.

Marine Protected Areas

MPAs can also provide a setting for pre-adaptation strategies, which are forms of adaptation aimed at protecting systems from projected changes. For example, reef-forming sponges, which appear to be dependent on grazing sea urchins for a strong structure, could lose ecosystem function entirely because the caged sea urchin will provide physical disturbance to seawater and counter the process of demineralization of the sea urchin (Chapt. 2). Of course, MPAs can also protect valuable and delicate marine life and ecosystems from the fallout of climate change, such as hypoxic and anoxic dead zones.

Marine Protected Areas (MPAs) can play a crucial role in mitigating the effects of climate change by protecting vulnerable habitats and allowing marine plant and animal populations to build resilience against those changes. They can also serve as valuable stocks of genetic and ecological diversity for the future, providing a rich "haystack" that contains "needles" for biodevelopment. MPAs, especially those that are "no-take, fully protected areas" without fishing or other resource extraction, also close off a carbon and nitrogen discharge pathway resulting from those activities (e.g., herbivorous fish fertilizer, benthic grazer impact, release of fixed nitrogen from sediments, and release of marine ammonia into the atmosphere when fish are processed). In some cases, overfishing or other pH-affecting forms of fishing can elevate pH, potentially acting as geoengineering.

CHAPTER 15

Health Impacts of Climate Change and Public Health

The health community has an important role in communicating with the enabled: when people perceive their personal health or the well-being of their families to be at risk from some environmental influence, they are generally willing to adjust their lives, often in cooperation with their neighbors, to combat the perceived risk. The health communities, at all levels where it has a presence, is therefore potentially a strong advocate for mitigation of the greenhouse effect. Would that this were a sufficient justification for action, in view of the clear imperative in health terms for action to be taken to reduce the quantities of greenhouse gases being added to the atmosphere. However, it is evident that permitting progressive escalating climatic disruption is not in humanity's best interest and that action to mitigate climate change is a priority in its own right. As public health becomes acutely aware of the implications for health of failing to act against the environmental abuse that has given the health community the role it adopts in relation to climate change, it becomes another strong voice to add to those already articulating similar arguments against delay in implementation of greenhouse gas management programs.

Climate change is already affecting people's health. Through extreme weather events, it is causing the loss of life, with consequent grief to individuals and distress to communities. Climate change is displacing populations, sometimes permanently, and the consequent political and social issues exacerbate other health issues. Malnutrition is, and will continue to be, exacerbated by reductions in the availability of the agricultural produce by which it is largely generated, and by changes in the public health context in which people live, work, and rest. Vector-borne and water-borne diseases are likely to increase in incidence, while inequalities in health both within and between countries are exacerbated by the effects of climate change.

Heatwave Preparedness and Response

Warming of just a few degrees in the summer months with a changed climate carries the likelihood that the healthy can become vulnerable, especially if the peer group operating from doors with bad intentions puts a temporary lid on the distribution of power supply that is adequate to meet peak temperatures. Heatwaves remain a critical threat to life unless nations exclusively rely on renewable energy supplies, consumption patterns change considerably, or exceptional foresight and planning is used to confront an increasingly dangerous summer threat.

Heatwaves are a regular occurrence in many countries, making the poor, elderly, and very young particularly vulnerable. Authorities preparing to avert high-injury temperatures need a robust plan, adequate shelters, and designated cooling centers that operate when the power supply is intact. The Red Cross has a plan for extensive door knocking to check on the welfare of the elderly. Recycling facilities that might catch fire, and coal bunkers that might ignite, have to be frequently and thoroughly monitored. In some regions, heat-waves are a top priority for the authorities because so many citizens die when temperatures spiral upwards.

Vector-Borne Disease Management

A second approach for the United States is to increase expenditures on research and development – not at universities, but by the units of the relevant state departments responsible for vector-borne diseases. These research laboratories should work on the strength of the vines so that excessive spraying is not required. This approach will be much more effective than broad-based work by academic laboratories that reflects normal priorities rather than the requirements of the state. Thirdly, the United States should support research in the countries in which vector-borne diseases are endemic. State aid to countries where dengue fever exists would support and enhance the economic development of such a country and not only protect the health of Americans, but also benefit quite a few million people in those countries. The actual dollars involved are in the range of $40 million, a negligible fraction of the cost of a new weapons system.

The wide variety of diseases in Table 15-1 includes a number of newly emerging diseases such as dengue fever and Eastern encephalitis. Such dissemination of new diseases is in large part due to global warming. Once a disease has started to disperse, it is important to develop good diagnostics. With a current virus spreading rapidly through California, I am appalled that the usual state agencies at the university say that they do not have the skill to search the required databases. It is astonishing that the state of California does not have the necessary diagnostic skills somewhere. In these cases, it is very important to act quickly. In the case of potential diseases not yet endemic in the country, it will be especially important to have reliable diagnostic tests available so that the disease can be exterminated before it starts to spread.

CHAPTER 16

Ethical Considerations in Climate Solutions

Other experts considered the ethics from both global energy and emissions policy needs and air capture and storage ethics as well as toward challenges facing the ethical professional toward significant scientific study on ethical/governance dimensions of climate change, including public policy choices. They concluded that ethical considerations of both mitigation and solar radiation management can be addressed. Society can increase awareness of potential dangers if mitigation cannot be sufficiently implemented or enhances public communication and upward help improve controls as well as support decisions, through careful experimentation of solar radiation management. Furthermore, as a result of a comprehensive American Physical Society climate study, ethical considerations regarding climate change such as obligations to future generations and atmospheric trust related to common atmospheric resources would help in legal claims. They were considered and reflect a robust body of social science literature.

When ethical considerations for climate solutions are discussed, the focus is typically on concepts such as moral hazard and equity. Especially since potential mitigation approaches range from cutting emissions to modifying the environment to using carbon capture and storage, the field of solutions brings new and challenging ethical considerations to the forefront. With these considerations in mind, the National Academies established bioethical principles to guide future decision making and also looked at trade-offs between implementing relatively straightforward reductions in carbon dioxide, from increased energy efficiency and retimed carbon emissions to approaches requiring cost and technological breakthroughs including some based on bioengineering, carbon sequestration, and other carbon negative approaches.

Environmental Justice and Equity

At the local level, the inequities are becoming equally apparent. Here, the affluent, protected locations of those in politically powerful urban areas. Residents of inner cities and those living in proximity to refineries and utility plants are far more likely to be exposed to hazardous toxic air pollutants than those in more politically powerful and well-heeled surroundings. And as climate change begins to hold the world's economies hostage, it is the world's poorest regions that are least likely to be able to deal with the problem. Efforts to correct the balance of this single-sided arm-twist have thus far focused both on the global level, where the UN Framework Convention on Climate Change has pledged $4.5 billion annually to attempt to globalize ahead of the problem, and on the local, where such major utility customers as the EPA have been directed to reduce their contributions.

A central aspect of the climate justice question is the inherent inequity of climate change impacts. While industrialized countries have produced the bulk of the greenhouse gases that are causing global warming, the global poor are expected to suffer the brunt of its consequences. As a result, the countries that contribute least to the problem and are least equipped to deal with it are the ones that stand to suffer the most. This is already happening. Climatic changes are affecting rain and snowfall patterns in already icy lands in the northernmost reaches of the globe to equatorial regions where farmers and forest-dwellers depend on the predictions of regular monsoons. Low-lying island communities in the tropical seas are already at risk from a rising sea level. Few see the costs and thus deprioritize what is rapidly becoming the world's most pressing problem.

Intersections with Human Rights

In 2003, the United Nations' former Secretary-General, Kofi Annan, bypassed the traditional tit-for-tat arguments concerning whether climate change as such is a human rights issue by stressing that protection against climate change must be seen as a fundamental human rights issue, including not just human rights, but rights in general. This type of framing is increasingly used in both conceptual and policy-making contexts, given the serious effects that climate change has on people's lives and well-being.

Intersectionality is a term used to argue that human rights violations and plight are so interconnected that no movement for their recognition should be seen as of greater import than another. Intersectionality is used to argue that humankind as a whole sets a high value on the recognition of human rights and that, in order to overcome violation, all rights should be given equal weight. That is not the way in which this term is being used in this section. Instead, this section is written against the backdrop of the interconnectedness of human rights problems, to bring the human rights message into the center of discussions on climate change and to work towards pulling climate change up to the top of human rights discussions.

CHAPTER 17

Technological Solutions for Climate Monitoring and

It is generally agreed that we can expect a warming of from 1 to 7 degrees Celsius over the next century from CO2 emissions alone, with other anthropogenic inputs tending to accelerate this resultant. Frequent explanations such as that the CO2 necessary to warm the earth an additional 1 degree Celsius by the year 2025 could be continuously supplied in terms of present resource usage by a small number of large electric utilities annually at a fraction of their gross income have prompted many citizens and their leaders to discuss the necessity and dimensions of comprehensive policy strategies to provide policymakers a full public benefit-cost analysis of greenhouse control measures.

Global monitoring and extensive data analysis are essential to this activity, and we must seize the opportunity to use new monitoring systems and advanced modeling and data assimilation techniques to improve our ability to predict future climate states. Such assessments require us to recognize and exploit the possible impact on society of foreseeable climate change and the so-called surprise climates that may alter economic growth. The cost of improving our research and development strategy is small in comparison to the cost of potentially large societal damage that inaction might allow. With leadership cooperation at the international level, this R&D effort should find more widespread acceptance. Although significant uncertainties exist due to the need for parameterizing the details of some physical and biological-climatological processes, there is a broad consensus among scientists that the net effect of rising greenhouse gases will be to warm and alter the atmospheric, ocean, and terrestrial environment.

Dramatic increases of atmospheric greenhouse gases, and especially carbon dioxide (CO2), are causing widespread and sometimes drastic changes in global climate with negative impacts on economies, environments, and national security. Recognizing this, climate research is in its early stages of becoming a focused, international, and systemic undertaking involving a range of new generations of satellite systems and advanced ground-based facilities. This work envisions their support of large projects, a heavy reliance on suborbital measurements for reasons of cost and flexibility, and a strong integration of science, engineering, and data management.

Satellite Remote Sensing Technologies

In this chapter, we assess the current and future capabilities of a couple of the most used instruments relevant to the mitigation of uncertainties using satellite remote sensing: the A-train satellites. Based on SWOT analysis and a survey of literature in the atmospheric and oceanic community, as well as our own funded and proposed NASA projects, we develop science needs for mission architectures, propose scientific applications, and discuss a polar orbit mission, the so-called "Climate Interdisciplinary Satellite" mission. Our primary goal is to examine how a space mission can better utilize these environmental great observatories to minimize uncertainties in climate forcings and climate feedbacks in the larger context of the global carbon cycle. What should be the long-term build-out for a constellation of new missions? Will these missions better assess uncertainties associated with AR4 climate stochastic properties? We explore how current missions observe, with future predictive capabilities, how to build strategic sustainable satellite constellations that focus on measuring the most uncertain relationships in a cost-effective manner. Our research is couched within the quantification of errors arising from satellite remote sensing of this nature. We provide a brief summary given mission design constraints and suggest a set of potential questions that the science team could potentially address to minimize such errors, were these designs to be implemented.

The last decade has seen a rapid increase in remote sensing technologies in space. These satellite remote sensing technologies have largely been dedicated to the scientific study of the earth-atmosphere-ocean system, particularly pertinent to the understanding of physical and dynamical processes and the understanding of the carbon cycle. Yet the last decade has also seen a significant leap in our capabilities of minimizing uncertainties in the understanding of the anthropogenic role in climate and climate feedback processes as well. These capabilities have become extremely important due to ambiguity in AR4 observations regarding the anthropogenic effect on the hydrological cycle, cloud feedbacks, biogeochemical carbon cycle feedbacks, and the understanding and quantification of weather and climate extremes.

Big Data Analytics for Climate Research

Tracking the changes in Earth's climate system at a macroscopic level through weather attributes such as temperature, pressure, wind, moisture, and so on over space-time scales has been the goal of climatologists for a long time. Since weather predictions are unreliable beyond a week or two, they resort to statistical analysis of weather station data either to understand the past or to build models for the future. Providing accurate and reliable predictions of temperature, rainfall, wind, and so on, even for the next season, let alone the next year, would be a huge challenge due to missing data, inadequate modeling of complex climatic processes, limited computational resources, and so on. This situation urged the meteorologists to either collaborate or look out for alternate analysis methods from different disciplines like applied mathematics, computer science, and data mining. Social scientists have been studying the socio-economic problems resulting due to climate change, but the impact of weather on these problems also needs to be factored in while doing the analysis.

Massive amounts of data are being collected through sensing and simulation instruments that are deployed over land, air, water, and space. This could lead us to the next scientific revolution, feel some. However, extracting valuable insights or knowledge from this data is a difficult and challenging task, such that conventional computational tools and resources cannot solve it. In this chapter, the authors have delved deeply into the large-scale big data analytics methods, tools, and techniques that could help the climate research community in understanding the climate change dynamics at a macroscopic level, which could in turn solve a myriad of problems from diverse domains like agriculture, ecology, biodiversity, energy management, smart and cool cities, and so on.

CHAPTER 18

Innovations in Climate Communication and Media Eng

The chapter explores the essential role of strategic communication in the climate change challenge. It reviews recent research on how best to engage diverse audiences. It suggests ways to improve communication about the private motivations and benefits of climate action. And it reviews innovations in public communication and the media. The result, it is hoped, is to give climate change communication the serious attention it warrants, both as a crucial policy tool and as a key public engagement and education strategy.

Innovative and strategic communication about climate change plays a critical role in explaining the causes and effects of global warming, shaping public attitudes and behaviors, inspiring policy action, and encouraging a spirit of cooperation among diverse audiences. This chapter will consider recent innovations in climate communication and media engagement and offer practical strategies for engaging the public around solutions to global warming.

Visual Storytelling for Climate Awareness

Our battles haven't really been on any real parts of our world, and they don't contain real tales about athletes, climbers, or heroes. Because climate disasters are our everyday lives. The human quest for pleasure, advantage and also the simplicity of pointing fingers includes conflicts. In order to create fictional images too far away from our ever-changing changing world, the luxury of having artistic works that spare spectators from everything they don't want to see. I have pictures of boys playing basketball in a new and rare neighborhood found on experimental photographs magazine. This exciting work is the consequence of a far longer, strong and intricate trip before those beings were overcome with acid-free strong and complicated support. Most of us stand ready to deceive oneself, while their families and people like us embrace the comfort of our narrative to the dreamed end.

The popular photo-sharing app, Instagram, can function as an influential medium to increase public climate change awareness—though increasingly frequent incidents of unusual weather are also convincing viewers. Now, many artists, advocates, and environmentalists use social media to share their interpretations of the scary scientific consensus. While it encourages the public to join the conversation, the goal of these works is to make people feel about climate change by gaining an understanding of the facts. Unfortunately, often the truth about climate change just isn't enough to provoke the action of people. We are full of people who are far more influenced by stories, who are engaged, who challenge themselves, who engage them and who are used to understanding and using stories of every scale and complexity.

Social Media Campaigns

Neither individuals nor firms are motivated to make sufficient conservation choices nor investments in energy-efficient capital because they do not fully internalize the cost of emissions. Inertia, lack of information, risk, and psychic distance are relevant barriers. Policies to overcome distortionary taxes, promote behavioral nudges, and manage the risks of discrimination will help lead behaviors to be in greater alignment with climate actions. Companies can play a role in stutter-stepping governments towards a reasonably robust economy-wide carbon pricing and/or emission-reduction policy package by reducing these stockholder and stakeholder barriers to effective climate action. Premium efficiency, demand response, green product demand, and supply chain efforts by firms support external policy advocacy and can generate real firm and global economic benefits, reducing risks of future business harm from climate policy action.

Despite increased awareness, the majority of the world's population does nothing to combat global warming. Experts and most informed individuals concur that extensive governmental action is required to reduce greenhouse gas emissions, but governments act most effectively when the public demands it. Consequently, we advocate social media campaigns that present arguments for governmental action in a format that is compelling, entertaining, highly informative, and in a location (such as websites and social networking groups) where people are willing to engage with the material. There is considerable evidence that such campaigns can cause behavior change and help build a pro-environment culture. Such a culture leads to public demonstrations and becomes a source of political pressure, which can influence government policy. Given the likely dangers of future damage from climate change, we believe that such campaigns are a good investment.

CHAPTER 19

Gender Perspectives in Climate Action

In most contexts, men have the greater relative control of power and energy in the home, which also may result in multiple benefits in terms of managing energy-related household expenses. These multiple benefits, in turn, may help to overcome male skepticism of renewable energy interventions or investments while reducing potential negative gender or social impacts. Women's functional needs and priorities also may not align with technology benefits and they may be bound by competing demands and time-constraints. A shift in gendered attitudes and perceptions on energy use, power dynamics, and decision-making is likely also necessary as renewable energy can influence both energy supply and use. Cross-sectoral partnerships with ministries addressing gender-specific development issues are therefore key to supporting women's empowerment and ensuring that a true understanding of the factors affecting equality as concerns the control over, social and political implications, and economic aspects of energy use becomes part of all policy discussions of climate and sustainable development.

Access to and control over resources, equal rights and responsibilities, and freedom from violence are essential elements, in addition to economic participation and influence in decision-making that can make climate and development policies more effective. In the context of renewable energy policy, programs, and project design and implementation, this means that it matters both how energy and energy services are used and who makes decisions about them. While issues of equity, participation, and decision-making have been considered in the design of renewable energy climate change policies, the importance of these issues, especially at the local level, tends to be overshadowed by a focus on large-scale technology deployment.

Women's Leadership in Environmental Movements

Women have taken leadership in the management of natural resources in their communities; many are the primary users. When development projects consider the needs of local women, then improved economic and environmental performance should be expected. I found an analysis of tenure legislation in countries in Latin America and Africa. I calculated (as a percent) that out of the total number of laws in a country, which allow women to hold title to land while still married, in their names alone. Percent of modern laws apply to both married and unmarried women. Percent of bank loans in the country are available to women as well as men. Data indicate that the more countries permit women to act independently and as economic individuals, the higher both the gross domestic product (GDP) per capita and the percentage of economically active women. This reveals that economic takeoff can happen in countries with traditional cultures and societies, where economic roles have not been modernized. Women are legally permitted to manage natural resources at the local level in many regions. Women lead campaigns to halt deforestation, to save areas of wilderness, and to advocate wider use of family planning in areas of rapid population pressure. Women have been leaders in addressing problems of local air quality and health. These movements encourage local decision making at the community level. They often find ways to transfer land from public to private ownership, thus recycling public subsidies originally used to promote development. This involves careful negotiation between community-based organizations and governments. In sum, powerful entrepreneurial behavior exists in organizations with collectivist values and goals. Women seem particularly adept in achieving this outcome, even if they do not want to be entrepreneurs and do not use profits for private consumption.

Gender-Responsive Climate Policies

The Paris Agreement prescribes a just and inclusive transition for people and places, and it is essential that future climate diplomacy develops national climate strategies that build equity, advance human rights, protect ecosystems, culture, and heritage, and embed consultation, ambition, transparency, and fairness.

Gender-blind climate policies have, in some cases, led to failures in both climate outcomes and effective economic and social development. The best way to make gender-responsive climate action is by listening to women's and feminist voices and supporting them through relevant gender expertise and by following the evolving and evidence-based guidance and frameworks.

Over the past decade, the Paris Agreement has led to a significant increase in finance flows to both climate adaptation and carbon mitigation, which are increasingly being delivered through international frameworks, as well as at the country level. These global and national plans and processes have varying results in addressing gender equality and the empowerment of women.

Gender equality and transformative climate action are both vital and interlinked. The SDGs cannot be fully achieved without addressing gender inequality, and global climate goals will not be reached unless gender issues are adequately addressed in climate action.

CHAPTER 20

The Role of Youth in Climate Advocacy and Activism

Educators must provide students with essential skills and knowledge to take action. Just as individuals who require the necessary information and skills to make practical, effective choices in their daily lives, students and young adults need insights and awareness to make meaningful choices about the U.S. role in international efforts to combat global climate change. With only 6.2% of the world's population, Americans already use half of the global output of refined fuels and roughly one-third of the planet's raw materials. Altered consumption patterns in the United States would significantly contribute to global climate change solutions. A broader curriculum can teach school- and college-age youth about the environmental impacts of their daily actions, the roles of informed citizens in solving environmentally thorny problems, and the experiences and viewpoints of people outside the United States.

Youth are poised to become transformative agents for climate action on three levels: the personal, the community, and the global. As individuals, young people are exhibiting their power as consumers by demanding to know the origins of the products they buy and boycotting the goods produced by companies with poor labor and environmental records. Young people around the world are participating in community efforts to build sustainable residences, create markets for organically grown produce, protect forests, and organize ridesharing. And at the global level, young people have emerged as valued spokespersons and agents of climate reform. Notably, students in over 50 U.S. cities now participate in an Urban EECO (Energy, Economy, Climate Opportunities) team that gives them a prominent role in city climate stabilization efforts. Local teams enable students to participate in the real work of building and maintaining a low-impact community and provide ways for youth to become leaders of community sustainability efforts.

Youth Climate Strikes and Demonstrations

In September 2018, people in over 1,000 cities in 100 countries joined the kids in their first Global Climate Strike. The turnout of these youths was like nothing the world had ever seen before. These well-coordinated and powerful strikes have continued to grow and inspire youth to stand up for their future and the future of the planet. So in festive resistance and support for the youth, individual adults should join in the youth strikes too. They are more fun and bring media attention to the youth's urgent and important message to politicians. This youth movement is having an extraordinary impact and is part of a worldwide wave of activism that is engulfing and inspiring people to find and fight for innovative and transformational change.

In August 2018, Greta Thunberg, a 15-year-old Swedish girl, started the first-ever school strike for climate in front of the Swedish Parliament building. She protested the government's inaction on climate and pledged to keep sitting out school until the parliamentary election, which was 3 weeks away. She posted a photo of herself with her sign on social media, and her courage and conviction inspired youth all over the world to replicate her school strike every Friday. Greta catalyzed what has become a world movement for the youth who are angry and frustrated about how previous generations have let them down and are ready to take action to protect their futures.

Youth-Led Climate Organizations

The questions that then arise for philanthropy are the usual ones in the face of the successful growth of a start-up. Will significant leadership turnover at the nonprofits involved occur? Might Power Shift institutions turn into secretariat-type organizations where students from diverse campuses come to meet? What will follow when the existing leadership "ages out"? I encourage philanthropy to carefully evaluate the group's capacity to sustain its work with less guiding effort from the current staff and leadership. If this step among others is accomplished, how can the existing organizations be more "institutionalized," so that their campaigns will continue to grow without student leaders requiring as much facilitation and mentoring?

Another substantial contributor to the climate movement are student-led groups. Power Shift, for instance, extended a 1999 effort among national student associations to focus on global warming. It then ignited the climate movement, supporting more than 4,000 youth in 350 campus groups in developing campaigns at 2,200 separate colleges and mobilizing students to press Congress and campuses countrywide. As part of an effort to make sure that coal-fired power plants were held accountable for meeting Clean Air Act restrictions from operating above state-of-the-art levels or violating prevention of significant deterioration provisions, Power Shift also initiated a series of lawsuits under the New Source Review section of the act. About ten more campus climate network efforts also exist, all started since 2006. The common cause of these efforts is building youth organizations capable of confronting climate change and seizing the opportunities to build a clean energy and green jobs economy.

CHAPTER 21

The Circular Economy and Resource Efficiency

Faced with these circumstances and also bearing in mind that the faltering of global accumulation is symptomatic of a systemic crisis that could transcend Republican interim presidencies, and hence the issue could be taken over by a government advocating international law in all its areas of competence, efforts have been made to design a strategy that will resolve the problem in a more or less traditional manner by issuing subsidies to stimulate the electrical economy conformity assessment industry so that climatic catastrophes do not affect workers. At first, it might seem that enacting excessively large strategic measures is paradoxical. However, as we have favored the privatization of security enforcement and have increased its inefficiency by destroying the social fabric as well as any reference to the substance of Western civilization, having reduced it to manipulation of power mechanisms alone, the key question being control of resource distribution, it is a fair hypothesis that subsequent normative shape will follow patterns previously described.

The issue of land degradation, understood generally as the loss of soil fertility, compaction of soils, moisture loss, salination, inadequate drainage, erosion, and runoff, has methodological repercussions in various international fora. As already mentioned, the EEA, in its 2005 overview of the issue, had already related soil degradation to that year's abnormal weather. In Spain itself, the MIMAD incorporation of soil erosion risk into CAP environmental conditionality is noteworthy, which requires Commission ratification. Nonetheless, observation has not yet been translated into action and universal legislation. Adhesion to the Stockholm Convention shows a lack of coherence relative to the catastrophic findings of the Stockholm Study, attributing the processes producing hazardous waste to the need for GDP growth. Indeed, similar considerations would apply to the barely plausible processes whereby one signed party after another keeps producing waste, while bearing in mind that there is a direct relationship to air pollution caused by fermenting CSG.

Waste Reduction Strategies

Innovative low-cost methods to cut household waste also produce climate-friendly lifestyle modifications since they lead to more responsible consumption. If manufacturers are made responsible for the entire lifecycle of their products, they will have an incentive to minimize packaging materials and use less toxic materials in their product. New regulations would make it more cost effective for manufacturers to design for recycling and reusing their products. Extended Producer Responsibility (EPR) is an alternative to traditional waste management and product stewardship that holds manufacturers responsible for the entire lifecycle of their products to prevent waste and damage to environmental resources. EPR means that the manufacturer is responsible for bringing the product to the end of a practical life and for recycling or eventually disposing of unused or worn out goods in a non-composite state. This strategy unlocks the value in items that would otherwise turn into waste. As the material and product has an economic value, in a well-functioning

EPR system the economic benefit outweighs the costs for recycling. These costs then become a normal operational cost.

Waste management is a significant source of greenhouse gases. Organic waste decomposes anaerobically (without oxygen) in landfills to produce methane, a potent greenhouse gas. Furthermore, incineration of waste produces carbon dioxide and also releases other pollutants into the air. Reducing the amount of materials we use or buy and consuming less contributes directly to efforts to minimize waste. Households can cut their waste in half by reducing the volume of waste they generate and composting their organic waste, effectively reducing these greenhouse gas emissions.

Product Life Cycle Analysis

Hitting the Greenhouse Targets: There are basically four swamps where technology hits the GHG targets: in new product and process developments that are triggered at the industry's own volition, outside intervention from governments that want to take some action or another, private sector activities that want to do something that others believe is a should, and public entrepreneurship and research and development facilitated by government and the military, both acting separately and in conjunction with the private sector. An industry reorganization theory known as path dependence tells us these are four different guns and they are aimed at different targets. Yet, when the shots finally ring through the cumulative air, they come together at different aggregation points in the cold fusion song.

To better understand opportunities and constraints, we need to look at the whole spectrum of product and process activities embodied in providing and using a product. In these classes of activities, no one activity can be simply and totally changed without affecting the others. Optimal levels of prevention and control are achieved through life cycle changes that balance these differing constraints and opportunities. To harmonize activities, the corporate board and others have reveled in policies and best management practices that seek to cross the functions, management levels, and refractory walls that exist within organizations. Such conceptual and institutional integration makes for real-world management information that is more inclusive and extensively connected.

CHAPTER 22

Cross-Sector Collaboration in Climate Innovation

At the same time, neither building nor transportation firms can engage in this development space because prototypes require meaningful beta testing. Under the present circumstances, it makes little environmental nor economic sense to reduce the potential benefits of the projects. In either case, each sector has strong incentives to engage in cross-sector collaboration. The economics of the hydrogen infrastructure represent the leading tangible example of incentives for this type of collaboration. Mobius suggests a tiered market structure for carbon credits. It is intended to help in the design of policy for carbon credits.

Many assume that climate issues are primarily related to the energy and transportation sectors. Because energy projects are capital intensive, leading firms in that sector often take years to develop beta, as well as actual physical prototypes. Frankly speaking, yet the intangible cash value of these projects often rises based on receiving carbon credits associated with climate and energy. Common knowledge is that policymakers have not acted to connect transportation emissions to the cost of building systems, whereas building improvement projects provide the most readily available way to reduce greenhouse gas emissions. As a result, there is no direct business case for sharing the risks and, ultimately, the profits of developing emissions-reducing technology. This situation is the primary barrier to innovation abroad. The gap between the potential of technology to mitigate climate change and the capital to facilitate it is huge.

Public-Private Partnerships

The wide variety of instruments discussed in previous chapters - a carbon charge, tradable permits, regulation, government procurement of technology, R&D support, government provision of basic public goods, and direct support for information and consumer purchasing - depend on an extensive programme of coordinated public spending and plausible administrative regulatory mechanisms. There are inherent limits to what governments can do to change incentives, especially in broader areas of social management that trigger many of our society's deepest divisions, such as populating the earth, rearing the young, enjoying lifestyles governed by concepts of "good taste", or encouraging idealism. Consequently, the earlier chapters of this book expressed greater confidence in the potential of public finance to stimulate a compelling energy future or to even arrest global climate change through the modification of these social divisions.

Academic-Industry Collaborations

To diffuse knowledge about relevant science and technology, companies could fund university workshops that address particular problems. By tapping into the broader set of appropriately skilled scientists and working across disciplinary boundaries, universities can provide companies with important breakthroughs and innovative solutions. One example of such a cooperative effort is the cooperative agreement that the German national telephone company, Siemens, recently announced with the German state of Baden-Wurttemberg. Given the high quality of the scientific work done by Baden-Wurttemberg universities in the relevant field of atmospheric physics, the company is matching the federal funds to establish a Weather and Climate Research Program with additional funding of up to 400 million marks annually for research in this area. Such a project is rarely unachievable on a smaller scale, in the United States or even among universities within an individual country. Within individual companies, funding for climate research is typically not the key issue. Instead, the problem with university cooperation frequently is to provide the necessary information and resources in a timely enough fashion to those in the company with problems.

Given the difficulty that most companies experience in understanding, much less keeping abreast of, the latest scientific research, why do companies not establish formal relationships with university researchers? One reason is that companies rarely, if ever, provide seed money or guarantees of funding in exchange for future discoveries, even while they grumble when academics, who need their support, seem to be frustratingly unpersuadable. Thus, a cooperative research activity is trying to create a no-strings-attached "intramural" fund to which companies could contribute at different levels, in exchange for a say in the formulation of problems. However, this approach, if successful, would meet only a small fraction of the full unmet need for funding of university research with industrial relevance.

CHAPTER 23

Cultural and Behavioral Shifts for Sustainable Lif

Recent modeling efforts suggest that the South will not be able to afford western levels of carbon-intensive welfare and that a sustainable energy future requires a cultural and behavioral shift. In response to these challenges, eco-cities, sustainable housing, lifestyle assessments, sharing services and time banks have been developed as an alternative material of semi- to post-consumer culture. In southern age-old religions can inspire modern approaches, whilst new organizational forms, such as tenancy-ownership or co-housing, offer sustainability in a networked cultural environment. However diverse these projects are, some common attitudes to cultural and behavioral shifts become apparent, that have the potential to shift both North and South towards sustainable living. Such synergy requires a multi-scale and multi-sector approach that draws on a variety of scientific disciplines.

This chapter presents only a small sampling of the initiatives happening globally that lead the way towards consumer-based sustainable living and Greenhouse Development Rights. The projects were chosen from Asia, Africa, Middle East, Latin America, North America, and Europe and concern sustainable consumption and energy conservation. Some will help shape the individualized, mass consumer culture of the future; others orient urban planning and personal mobility; some offer a critique of work and leisure time; and all of them point to innovative hybrid solutions that mix western and southern traditions in sustainable living.

Minimalism and Conscious Consumption

Minimization brings many advantages and allows us to avoid the goods that contribute to environmental degradation. While some damage caused by overconsumption cannot be avoided, minimizing our consumption can help reduce the amount of waste we produce and the negative impact on the environment. People can strive for conservation, but it is important to recognize that any type of good, unless it can be regenerated by nature in the same proportion it was extracted, will have some environmental impact. This is because disposable goods lose their value of environmental sustainability. However, by minimizing our consumption, we can avoid objects that contribute to degenerative consumption behaviors, not only at the moment of enjoyment, but also in the production and possession stages. Minimization is not just about owning fewer possessions, as it can also be driven by a desire for quantity (the one who needs less is the richest) and a willingness to avoid objects that promote degenerative consumption behaviors. When we prioritize minimization, the social ego associated with consuming fashionable objects becomes irrelevant.

People start to consider changing their values of life when they are overwhelmed with excess and understand that excessive consumption of material goods does not bring happiness. This process can lead to a shift from mindless consumption to conscious consumption. For many people, happiness does not come from impulsive buying, but from being mindful of what they consume and avoiding wasteful and degenerative behaviors. As the literature shows, people do not seek goods or happiness in and of themselves, but rather emotional states that can be achieved through the use of goods. Instead of chasing after extravagant goods that provide immediate gratification, they choose to minimize their consumption and focus on what truly brings them happiness, such as free time and personal fulfillment.

Local and Seasonal Food Movements

In order for the local and seasonal food movements to continue to grow, develop, and mature, they will need to establish, maintain, and develop infrastructure in order to make it easier for consumers to obtain a wide variety of fresh and preserved locally produced foods. In order to tear down barriers to obtaining market access, local and regional entrepreneurs will also need to harmonize and form alliances with each other for the purposes of aggregating customer bases and making it easier for producers to reach the consumer. Regaining competitive edge will necessitate technologies and organizations that can provide innovative ideas and knowledge to improve the ability of local food production systems to capture markets and create incomes while being statutory creatures that play by statutory rules. At the same time, the focus on the competitive edge of local food production and automation should complement exceptional managerial skills, ethics of doing the right and fair thing in production, and fostering strong local economies and communities.

Support for the concept of producing and consuming locally grown food has been energetic and growing across the U.S. Local food has a significant and growing market share. The primary objectives of the local food movement include increasing the visibility and trustworthiness of the supply chains of the food one consumes on a daily basis, and increasing the availability of locally produced food. The objective to obtain the freshest and ripest food is one of the prime motivators behind the purchase of locally produced food. Other motivators include concern about environmental issues commonly associated with large-scale monoculture or industrial crop production, as well as economic and rural development issues, the need to support small and mid-sized multi-generational farms, and the need for increased or improved agricultural infrastructure and increased agricultural-related employment.

CHAPTER 24

Legal Frameworks for Climate Action and Environmen

Given the degree of reported dissent regarding environmentally conservative actions, including market-based solutions to environmental protection, what legal frameworks are available to deal with the threat of climate change? This is the propitious time to consider such an issue given the 2020 election season and the competing platforms of major and minor political parties alike. On the left and the right, environmental protection is being debated extensively, and the competition among environmentally friendly solutions regarding assertive implementation is unique. Nonetheless, regardless of the success of these maverick proposals, other frameworks are available and can be utilized without considerable delay or the need for legal reform. This chapter reviews the most important legal frameworks used to restrict carbon emissions. High school Civics courses fail to adequately emphasize the number of U.S. governments, a hierarchy that suits the complexity of environmental challenges quite well. Any level of government can impose greenhouse gas regulations. This chapter collectively reviews this regulation by discussing the most important environmental laws, with the last section briefly discussing the roles of non-governmental organizations and non-governmental systems of environmental protection.

The common images of environmental laws are regulations dictating the chemical constituents of human effluent placed into the environment, such as the air, water, and soil. But a growing number of laws impose such mandates on "greenhouse gases," such as carbon dioxide and methane, produced often through the generation of electricity, transport, and industrial operations. While less percipient regarding the threat of climate change, older laws were enacted prior to the recognition of the severely damaging consequences of unconstrained carbon emissions. Some laws indeed encourage carbon emissions, for instance, by providing favorable tax treatment for carbon consumption (think about the carbon content of gasoline and the related tax). Over the recent years, policy discussions have sought to reverse such carbon promotion by advocating regulation, and ideally utilizing the benefits of a market for these externalities generated by excessive carbon generation. Market-based solutions have twelve years of experience in the United States though the Regional Greenhouse Gas Initiative's implementation in New England, but date their origins far earlier with the sulfur dioxide market slighting in the mid-1990s to address acid rain. Market consideration was discussed much earlier, but logistical and political hurdles delayed its implementation.

International Environmental Agreements

Since their entry into common use with the 1981 publication of the 1981 Premiers and the 1983 Coalition treaty, the term "international environmental agreement" (IEA), nearly always used in the plural, has come to refer to a class of agreements. These are international in the sense that they include three or more parties, and they address biological, geological, or atmospheric phenomena observed at the Earth's surface. The phenomena may be linked (e.g., greenhouse gas accumulation, ozone thinning and resultant radiation exposures, fisheries' strain and future depletions) or independent (e.g., preservation of biological or inanimate repositories). The class includes both global and regional environmental agreements. These are geographically defined as touching all land surfaces and the contiguous waters largest to the Earth's polar regions, or some part thereof. The agreements vary in scope, ranging from those concerned with a single problem, such as oil pollution, to those addressing environmental issues in their totality. They differ in complexity, functioning, and effectiveness. They have succeeded in treating to some degree the environmental problems which prompted them on two-thirds of the occasions in which parties sought to commit themselves.

National Climate Legislation

Quality tests provide a valid vehicle for the control of pollution not only where a technology ("filter") can be mandated, but also where less stringent performance-based goals ("standards"). Tradable permits do an effective job in controlling sulfur emissions. Tradable permits for carbon may be part of any post-Kyoto strategy. Tradable permits are important "because, first, they are mandated by law, and when laws are passed and change in response to changed circumstances, the rules of the game for ordinary citizens are known. Second, they can be priced with reasonable certainty. Third, compelling alternative strategies are as-yet unproven. Fourth, tradable pollution permits ensure that each unit of capital is used in the least-cost manner to reduce pollution."

The failure of the world community (with the notable exceptions of the United Kingdom, Australia, Estonia, and not many others) to seriously address climate change has left many environmentalists in despair and has led some researchers to bypass their national policymakers and seek to foster a change in the path of economic and technological development. These efforts, notable as they are, do not remove the need for national policymakers to become engaged in a responsible and thoughtful manner. What policy shifts would be needed to impart a new direction to the current unsustainable path? It is easy to view national policy shifts in the United States in the context of this nation's failure to ratify the 1997 Kyoto Protocol. Yet laws implementing U.S. policies to address climate change are being enacted province by province, state by state, and at the federal level. These uncoordinated laws and regulations can have very significant effects on carbon emissions.

CHAPTER 25

Inclusive and Equitable Climate Solutions

Economic theory offers several principles that can be found useful in developing an understanding of environmentally motivated voluntary contributions and seat-belt-wearing behavior. These indicate that a bottom-up approach, focused on individual behavior and corporate strategy, can be used to augment top-down economic incentives and government policy as the economy adjusts to the growing concern over global climate change. People often fail to internalize the external costs they impose through their behavior. Yet, historical data on past environmental behavior indicates that voluntary actions can alter this path of adverse consequences. Policymakers should therefore seek to promote a wide variety of complementary climate change activities.

For the most part, proposed or expected responses to global warming are conducted at a high level with little direct consideration given to the individual and societal factors at play. Different behaviors and constraints affect how emissions are produced in the first place, and how damage is avoided or controlled. As we change the incentives or the speed at which adjustments are made, we can also change the expected conclusions that come from economic theory or modeling. Different incentives or expectations for behavior should also influence the construction of a response to global climate change.

Ensuring Access for Vulnerable Populations

It is important to ensure that responses to current and future climate-related public health threats do not disproportionately favor the rich, well-educated, and technologically advanced populations of the world. While advanced therapies and vaccines are an important part of many solutions, the initial, most cost-effective responses to many public health threats - be they related to climate change or not - are often the simplest, most straightforward approaches that respect local culture and tradition and do not rely on complex, expensive technologies. Providing international assistance to identify current and emerging public health threats, basic support of public health functions helps ensure that benefits are shared by impoverished countries or communities of the world and maintains a safer, more disease-free world for all of us.

As with many public health issues, the populations most vulnerable to extreme weather events and increased infectious disease and malnutrition are also the least likely to have ready access to medical care. People living in impoverished nations, poorly developed communities, or remote locations are more likely to live in homes lacking proper windows, doors, and screens - common hosts to disease-carrying insects like mosquitoes. Many countries are hard-pressed to provide immunizations against common, potentially life-threatening diseases, and few of these countries have the resources or infrastructure necessary to develop a surveillance and rapid response system to the changing patterns of infectious diseases that will emerge with climate change. Furthermore, poor or remote communities are particularly vulnerable to accidents, trauma, and emergencies brought on by extreme weather events.

Addressing Climate Injustices

Energy bills are not being addressed in affordable housing programs, but cash flow and budget analysis and projections and conservative loan counseling could often detect this problem that will not go away. Social security recognizes that low-income seniors have an increasing burden because of energy inflation. Four years ago, there was a cost of living increase in social security payouts for the first time in five years. The need is even greater now with energy prices more likely to go up than down in the months and years ahead. And the burden of higher energy prices will become more crushing as the months ahead increase the income of those most likely to be affected.

Many households are already spending more than they can afford to heat them - using 14 percent of income for utility bills. There are several factors that make the energy burden affecting low-income households different from that of other groups. These include the income level of low-income households and the fact that low-income households do not have the resources to make the long-term investment in energy efficiency that could immediately lower their bills and reduce the energy burden while benefiting them for the life of the measures. The possible long-term loan offers no solution here. Low-income households are often in rental housing rather than owning the house.

CHAPTER 26

Technological Breakthroughs in Renewable Energy St

We were amazed by the passion and high quality of the scientists and executives that put in common an impressive and exciting sum of innovative ideas and cutting-edge researches about electricity storage, which has long been acknowledged as one of the main bottlenecks of a fully sustainable and reliable power system. Cost-effective, democratized, intelligent, and optimized energy storage associated with distributed, PV-energized, strongly interconnected microgrids can drive a new wave of technological, economic, social, demographic, and geopolitical breakthroughs on a global scale.

Last February, I hosted a conference in Washington, sponsored by the House Renewable Energy and Energy Efficiency Caucus, which I co-founded, on a new technological breakthrough that could be a game changer. Dropping steadily costs of renewable sources - characterized by increasingly softer costs of solar photovoltaics - and their soon-to-become-competitive costs all over the world are breathtaking and revealing in terms of a unique moment in history. At the time of such triumphalism in the wind and solar power industries, it is also important to think about the two main bottlenecks that can slow down their growth: their intermittency and their predictability.

Battery Innovations and Energy Storage Solutions

The current Tesla Model S suite has 7,000 lithium-ion batteries that power the engine. Those 7,000 power sources supply the car with enough energy for a 265-mile range. For every unit of battery, the 265-mile range costs $100. In comparison, the gasoline-powered car rooming in on the Model S has a 300-mile range and only needs a $10 tank of gasoline to refuel. The electric car also takes half an hour to fully recharge at any Supercharger stations. These are the basic problems with the electric car, and these are problems that solar-powered cars need to overcome. The performance of a lithium-ion battery is determined by the ions moving between the electrodes. Current supercapacitor designs have limited energy with the safety concern that is inherent in high temperatures. The silicon anode abandoned product is able to charge the lithium-ion battery. This new product sacrificed life for a significantly shorter recharge time. When designing new products, engineers need a cathode design that has the increased lifespan of a conventional cathode. While supercapacitors help provide acceleration, and capacitors have helped reduce charge times, a new battery must improve the energy density of the design from 200 watt-hours per kilogram to any number which is greater than 300 watt-hours per kilogram. Scientists at Stanford University think they have found an answer to that solution.

Electric cars have a battery problem. The number of electric cars on global roads is growing every year. We are still in the early stages of electric car development, but 600,000 electric cars are produced every year with the market doubling every year. While that growth is very impressive, it is a drop in the gasoline-powered ocean. The

600,000 electric cars produced in one year are outnumbered by the number of cars on global roads produced in one day. Electric vehicles still only represent an incredibly small percentage of the total car market. One primary reason for this fact is the price of the battery. The battery is the most expensive component found in electric vehicles and electric vehicles have come about because of battery cost reductions.

Grid Modernization Technologies

Smart Grid can be defined in many ways based on the needs, interests, business models, and technologies of its observers. This paper seeks to describe a comprehensive view of the Smart Grid. First, we provide an operational definition of the Smart Grid that modelers can use to simulate its cost benefits. Using this definition, we operationalize the many potential benefits of the Smart Grid that technology advocates express. With these benefits, we can then evaluate the potential costs and benefits and equity implications of the Smart Grid enabled by policies, system conditions, and technology. The operational definition of the Smart Grid we provide provides general guidance to researchers and appraisers of Smart Grid deployment opportunities. While this paper is neither a comprehensive review of the literature in support of Smart Grid technology deployment nor a comprehensive boundary spanning, technology, or market analysis, it provides a rich foundation for Smart Grid planning processes to leap forward. With this paper's unique foundation, planners can leap forward to plan and shape an inclusive operational Smart Grid definition.

Grid modernization technologies add some important tools to the portfolio of greenhouse gas emission-reducing technologies for the electricity sector. Grid modernization concepts and systems may be incorporated into the "Smart Grid" of the future that will cost effectively and robustly simultaneously enable the integration of renewable energy; support the following in two-way flow of electricity and information between the utility and its customers; provide system reliability and security; and reduce greenhouse gas and criteria pollutants emissions. Grid modernization technologies utilize elements that seem mostly familiar. Electricity metering and billing infrastructure must be capable of recognizing and billing users for the quality of power they are using and when they are using it. System components such as storage must provide as well as store power. Customers at the end of the lines must be more informed about the relationship of their actions to the costs they impose on society as well as be able to communicate and execute those actions.

CHAPTER 27

The Role of Philanthropy in Climate Solutions

But let's break through that modesty and examine how big philanthropy can change the world. In film or management studies, artists immortalize the power of Dogooder, Lady Bountiful, or George B. Chadwick as indirect saviors of community, lost cause, or dispossessed. That history has not ended. Philanthropy still has the power, collective and individual, to trigger new solutions and mobilize vast fields of endeavor. The secret advantage of philanthropic action, unlike government fiat, state subsidies, or global interconnections, is its ability to act without fear, cooptation, or delay. It exists between the undertow of cable news and the bureaucracy of large government programs. The philanthropist, more nimble than the state, is much less disciplined. Unlike the bureaucrat, creativity and energy replace forms and directives.

Although philanthropic donations to help solve the climate crisis are at a historical high, the sum is but a minute drop in a vast global bucket of investment and action. Public support for all charitable giving across the globe currently stands at near a collective $75 billion. This support, impressive as it is, has its limitations. And yet—despite these limitations, in the vast universe of how philanthropy might tackle global warning, the philanthropist should confidently aim for the planets. It is true that some philanthropy will miss its target—and indeed, many philanthropists purposely turn away from addressing the overarching planetary problems, preferring the more immediate comfort of addressing local problems within direct control of their philanthropy.

Foundations and Climate Change Funding

With only existing technologies, it is difficult to see how actual greenhouse gas emissions can be substantially reduced. Substitute alternative energy supplies too soon, and we lose large numbers of species. Public funding pays for two methods and provides subsidies for a few others - which limits diversity and potential job creation. Troubled countries suffer economically from large carbon emissions as the United States generally benefits. Only when Americans understand our role as part of the global community will they see and acknowledge the morally corrupt nature of our shortsighted decisions, and the need to preserve economic and ecological balance with considering new possibilities.

The importance of healthy ecosystems and economic development to ameliorate poverty and other global problems is largely underestimated. Too often, other global warming solutions involve technological extreme makeovers or regulations that provide only partial help. We have no right to ignore the potential of more ecologically based possibilities available to us. Excluding an integrated solution limits the opportunities for slowing and reversing devastating effects. We must change our approach before we and many other life forms fall over a global warming world precipice - investing in a richer, more ecologically viable future is our only choice.

Impact Investing in Climate Initiatives

The salient questions for profitable clean technology investments are always the same as for any other kind of investment. Is there a solid business model? What is the return potential for investors? Can the company deliver an energy-saving product that meets currently or soon-to-be existing regulatory and marketplace requirements? Are there any competitive advantages (technology, brand, distribution, and the like) that can be expected to endure? Some of the big players in the business world are listening. For example, according to projections by market research firm Clean Edge, Citigroup and Bank of America are among several corporate giants positioned to invest significantly in clean technology in the next ten years; Microsoft, it is estimated, will invest $100 million annually in energy efficiency.

The urgent global need for action to reduce carbon emissions presents an equally urgent opportunity for social investors to strengthen our commitment to social and environmental benefits, including preservation and enhancement of natural resources and addressing key danger areas like global warming. A 2006-2007 report by the Stern Review on the economics of climate change for the British government stated that reducing carbon emissions costs about 1 to 2 percent of global GDP annually, while inaction could reduce future global GDP by at least 5 percent and perhaps as much as 20 percent a year - a risk of prosperity reduction that should concern and interest every social investor. Two key areas where social investors are devoting close attention are corporate green investments and climate modeling. These themes can be seen as areas of shareholder activism, as corporations adjust to a changed global environment, and as opportunities for venture capitalists and other private investors to fund new businesses, technologies, and products.

CHAPTER 28

Art and Creativity in Climate Advocacy

It is not a coincidence that recently developed sustainability projects increasingly rely on art. To revisit the Ultra-violet contest, you will see its success together with the appeal and the artistic message of the mascots. Public spaces have been used for projecting short films, whose powerful message about the vulnerability of the environment has conveyed to the public the meaning of coastal resilience. In another recent example, a large mural in the capital of Sint Eustatius island in the Caribbean sends a call-to-action on using renewable energy, whereas an art education program impacts the children and students of the same island. The Climate Advocate Platform (CLARA) brings together 32 groups involved in theatre, comics, and illustration, as well as in climate protection and raising awareness. The recent awareness campaign of Oceanografic therefore included works of art produced by local artists. Artistic creativity is the communication secret of projects that bring together aquariums, artists, and sustainability. Expressive art has also become an expression of the strengthening of community ties in rural unserved areas. And, at the site location for intensive political agenda discussions, art represents a major resource for tourist attractions. It seems that the ceremony elephant in the room, the lobbying activity, is no exception.

The author had not originally intended to cover artistic and cultural activities in this report. His deep respect for art, together with the title of this section, "Art and Creativity in Climate Advocacy," compelled him to write about this subject as well. The work of renowned climate activists portrayed next to renderings of the 12 winning submissions of the Ultra-violet mascots design contest and masked people performing on a stage in Ibiza show that art does have a role.

Environmental Art Installations

The Hall of Grids and Mirrors is a recent site-specific installation made by artist Elise Goldstein. Located in the Bronx, the purpose of the Hall is to "remake the usual flow of movement and intrigue the walker to contemplate time and space within this interchange." Hall of Grids and Mirrors opened on September 12th and is part of the New York Arts in the Parks Program presented by the New York City Department of Parks & Recreation.

On view will be a portable alcove room, hung with arrays of reflective Mylar panels, a continuous wall of grids, and a free-standing 10-foot-by-10-foot grid in the center aisle. Grid and mirror configurations will allow the viewer to see in real time their own image into a potential infinity.

For the viewer, the Hall of Grids is disorienting, but the mathematical precision of the infinite reflections is intended to simulate the feeling of Hubble telescope images or scientific nano images. Sponsored by the New York City Department of Cultural Affairs and the Bronx Council on the Arts. This year, the Bronx Celebrity Series theme is a Celebration of Women in the Arts.

Music and Performance for Climate Awareness

It is very important to find new ways to communicate key climate issues to different groups of people in different regions. Today's science is also global. Physicists, chemists, and other natural scientists cooperate with colleagues from all over the world. They also work with computer scientists and engineers. Yet the main forums to communicate the latest scientific results are science journals and international scientific conferences. Scientists reach the wider public via the media. To reach the wider public with discussions about important global issues such as climate change is more difficult. On the other hand, it is an increasingly urgent responsibility. Far too few people know what science has to say about climate change and what we should do about it. We need to change that. The fight against climate change is an essential part of the world's development and international scientific cooperation is its condition. Science museums could be a great help in explaining the things that scientists believe are most important.

While our quest was going on, our Swedish friends were busy with their own musical performances. Throughout the United Nations Climate Conference, the Swedish Embassy in Buenos Aires organized a series of World to the People performances. The musical events showed the importance of the Paris Agreement and brought the message of the urgency of addressing climate change to a wider audience. In her welcome speech at the first of these events, the ambassador said, "We hold this event today as we believe it is necessary to put the spotlight on climate change. Our engagement in climate issues is determined by our conviction that these issues need more attention, more effort, and more public support. That support can only be won if the population is aware of what is at stake." Körklang sang a cappella, and the talented musician and composer, Anders Hagberg performed solo. The Sound Kitchen was part of the mantle of artists who performed in the embassy garden. The range of musical performances showed that music and the arts can be potent vehicles for delivering important messages.

CHAPTER 29

Psychological and Behavioral Insights in Climate C

This article in The Psychology of Climate Change describes some areas in which psychological and behavioral research can help in enhancing climate change communication strategies and ultimately in stimulating effective climate change response. In general, there are many possible insights from psychological and behavioral research that can inform interventions to support climate change response, whether they involve broad-based public efforts to reduce greenhouse gas (GHG) emissions or interpersonal support for lower-carbon lifestyles. We consider several important topics of climate change communication where psychological and behavioral perspective are not.

This article reviews research in the area of climate change communication, with particular emphasis on research that incorporates behavioral and psychological insights that can help individuals and societies understand and confront climate change. The policy implications of this body of knowledge are framed in the context of addressing climate change from the perspective of the planetary emergency in which appropriately scaled climate policy (that is, aggressive, justice-oriented, and urgent) is crucial if humanity is to avoid overly severe climatic change. This goal of planetary emergency response suggests several avenues for research and climate communication that moralize and focus on the effects of climate change, make the impacts of climate change more concrete, and treat climate change more prominently and less noxiously.

Behavioral Nudges for Sustainable Choices

As a nudge example, consider a very simple social marketing program. In college, it was a campus policy to emphasize the wording on the portion of students' mailings that typically contained environmentally positive messages. Specifically, the seed of choice would include a phrase such as "Be Part of the Solution." Even more memorable for the students was the similar close: "Don't Be Part of the Problem." Attaching these phrases at the bottom margin led to a nearly three-fold increase in student response rates, presumably because they wanted to be part of the solution rather than part of the problem. While these phrases are usually partnered with a hard-to-predict event, the potential benefit is clear: more student responses mean more environmentally conscious students. These students will expect Hillel to be more responsive to their needs, as more activities are offered for the same price, the same planning grants go farther, and the same facility can be improved sooner. The Hillel will benefit from having attracted more of the solution.

Perhaps the most notable example of a green nudge at a macro level comes from energy-efficiency labels on appliances. Instead of the classic nudging example of automatically enrolling government employees into retirement savings accounts (as a way to encourage saving), or requiring that colleges list healthiest foods first in cafeterias (as a way to promote anti-obesity behavior), certain countries really encourage purchasing of energy-efficient appliances by labeling them in a particular way. Such labels, such as those found on qualifying ENERGY STAR products, did not dictate that companies stop making energy-inefficient products or that consumers stop buying them. However, they give both types of individuals (uninformed and informed parties) the information needed to think about the consequence of their actions. In general, these programs are thought to result in net social benefits. Domestic energy bills go down and pollutants are reduced, resulting in health impacts and greenhouse gases lowered. However, if the market were perfect, and both the buyers and the sellers were fully informed, then these effects could be completely realized and regulated. All imperfections, be they information, pricing, behavioral, social, or other types, can be addressed with specific policies, such as nudging.

Society is complex, and people will not always make the choices that best meet their needs. To address this, economists and psychologists have discovered that policy proposals called "nudges," which influence people's decisions without mandating behavior, can greatly increase the likelihood that we achieve our desired societal goals. Thaler and Sunstein (2008) elaborate on this social behavior in their book, "Nudge: Improving Decisions About Health, Wealth, and Happiness," and discuss policies that "nudge" individuals toward specific courses of action. Tools for changing our behaviors and guiding us into better decisions have considerable promise for sustainable solutions. As each of our daily actions sends signals into our markets and to the environment, these signals force us to ask what those signals are saying about the desirability of our acts.

Cognitive Biases in Environmental Decision-Making

However, the robustness of behavioral economics results and the sheer volume of repetition over the past three decades make the motivation for conducting experiments particularly in individual subfields redundant. Indeed, a short reflection on modern economic problems easily leads to the conclusion that Homo economicus is not an adequate model of economic behavior and that many experimental results make perfect sense in the context of what Daniel Kahneman and Amos Tversky had proposed in the 1970s under the prospect theory label.

The question of why a motivationally autonomous actor would consistently make harmful decisions impacts was, in large part, answered in behavioral economics. Over the past decades, many studies have demonstrated that individuals regularly fail to accurately perceive and evaluate climate risks. The initial response to what seemed like counterintuitive results was skepticism. Why would people not feel threatened by rising sea levels, food markets hit by droughts, winters without skiable snow in Europe and the United States, or sweltering temperatures in India and China? Moreover, experimental economists typically want to ensure that participants in their games make rational decisions. A variety of games have been developed to show that Homo economicus predicts and ultimately obtains the correct results. If behavioral anomalies are observed, this undermines the realism of applied economics results.

CHAPTER 30

The Circular Economy and Sustainable Fashion

The global textile industry contributes to 8.1% of CO_2 emissions, very close to the 8.9% of the carbon footprint of the mining extractive industry and half that of the building and construction sector. The goal of zero carbon may support other strategic goals, such as the modernisation of industries, clean energy and smart sectors, operational efficiency and sustainability, new research, new economic opportunities, breaking down social barriers, maintaining biodiversity and ensuring a better life; all providing a healthier and cleaner environment. The shortfall of scientific knowledge is balanced against the importance of this sector to the economy and by its primary function of providing us cover in clothing and interior decor to help live our lives.

The fashion textile industry is a major contributor to global warming due to the high levels of CO_2 emissions throughout the complex production supply chain. The sector generates difficult to recycle waste products following temporary use. Fast fashion also encourages frequent consumer visits to landfill sites, increasingly making fashion an out of time hobby rather than a catalyst for innovation and creativity. The paper describes some of the structures and patterns contributing to the current unsustainable situation in the fashion textile industry. It outlines some research areas in the field of textiles that can act as enablers and describes some potential solutions using scientific and managerial techniques towards a more circular and sustainable economy including proposals for encouraging the fast fashion consumer. The proposed sustainable solutions appeal for a redesign of the slow fashion concept in a more ethical way by targeting individual desires rather than using the fast fashion model.

Upcycling and Repurposing in Fashion Design

By applying upcycling and repurposing concepts to fashion textiles, not only adds value to the materials but, applied at an industry level, assists in reducing the quantity of textile waste that ends up in landfill. As a practice, fashion's use of textile resources and unsustainable lifecycle directly challenges many ecological systems. But with the adoption of upcycling and repurposing by fashion designers within the mainstream, rapid evolution of low-impact product options with a high turnaround in the marketplace is enabled. Consumers are then offered opportunities, beyond wearing the clothes (shoes, accessories), to participate more directly in a fashion-led, resource-friendly recycling activity.

Although eco-friendly fashion can be relatively expensive, adopted notions of upcycling and repurposing practices in fashion offer consumers an opportunity to participate indirectly in the movement through their purchases of goods made using similar concepts. Research into both upcycling and repurposing as a specific design strategy in the realm of fashion design offers industry practitioners a design approach and can form the basis for the design of new products while reusing textiles that would otherwise be discarded as waste. In applying upcycling and repurposing within mainstream fashion, textile waste becomes a valuable resource from which to develop fashion-led solutions.

Eco-Friendly Textile Innovations

The textile industry is beginning to make strides towards a circular economy, but we have much further to go. An emerging textile strategy developed involves creating clothes that can enhance human health by reducing pathogens' risk, protecting against electromagnetic interference, or adjusting body temperature with textile technologies. Combination of these and other fashion products with design for circularity foster innovations and stimulate behavioral changes in relation to both designing and using clothes, which can drive great social, environmental, and economic progress. Companies such as Adidas, H&M, and Burberry are currently researching biodegradable and recycled materials to produce clothes and accessories at the mainstream level.

Rising affluence and consumerism in both developed and developing countries have contributed to the epidemic of textile and clothing waste, as fast fashion becomes cheaper and more disposable. According to an report, 73% of the 53 million tons of fiber that the fashion industry uses to make clothes is landfilled, incinerated, or recirculated as lower value material every year. Fashion's textiles, with synthetic polymer or semi-synthetic polymer as the primary fiber, are not only in the necessity of design for recycling and biodegradability, but also entailed to make clothes' pollution in the scenarios of wearing and laundry through technology-enabled.

CHAPTER 31

The Future of Climate Innovation and Emerging Tech

Many popular methods for estimating the costs of reducing greenhouse gases have inherent limitations, underestimating societal benefits during the process. The economics of climate change has only recently begun to investigate in some detail the types of innovations that may result from efforts to curb greenhouse gases. Innovative responses to lowering the cost of abatement can augment the power of these more traditional methods. As inspired by Gates, much more focus in contemporary discussions about climate innovation should be on "jumping" to clean technologies, especially those that can be rapidly scaled. Building a low-carbon economy means focusing on the same kinds of expansion in technological progress that have fueled the advances made so far, while learning to better anticipate the social cost of the carbon that is emitted. By understanding the role of innovation in shaping present and future climates, better decisions may result. Small improvements in those facts could lead to enormous research and development (R&D) dividends.

The challenges of global warming and climate change are among the most pressing issues today. They are also more complex, enduring, and multi-faceted than many other problems, especially in the effort required to combat them. The myriad and growing complexities of climate change demand a wide range of potential solutions. Innovations, if properly incentivized and encouraged, can help to find some of the most effective potential solutions. Developing new technologies for climate mitigation and adaptation may be costly, but they are also critical. It is important to recognize that future climate-friendly innovations can help to solve the biggest problems with many people, providing opportunities to grow the economy in so doing. As Pascal Canfin and Al Gore recently stated, investing in innovations to curb climate change can actually help the economic recovery. The uncertainty, complexity, and range of options for responding to climate change means that the variety of solutions that have been suggested should absolutely be embraced. It is also undoubtedly important to identify, understand, and expand the best of these potential options.

Artificial Intelligence in Climate Modeling

At the highest resolution, global atmosphere models calculate physical properties of the atmosphere for many altitudes across the world. These variables include average air temperature, average air pressure, winds, and average water vapor and precipitation. This list is not exhaustive and tends to change as more is learned about processes thought to be important in shaping clouds, the variable most directly linked to the Earth's reflective properties. When focusing on just a single aspect of the physical world, researchers can increase the number of models globally so that other aspects, such as the number of clouds and their reflectivity, are sufficiently resolved to enable researchers to draw robust conclusions.

Models used to simulate future states of the climate system are much more complex than the models we use to forecast tomorrow's weather. One climate model does not exist on a single computer. Instead, it is a suite of hundreds of software modules, each modeling a different component of the Earth system – its atmosphere, ocean, ice, and land. Only by understanding the interactions between these different components can we hope to answer the questions of how the climate might respond to, for example, a doubling of atmospheric CO2 rising global temperatures by 2° to 5°C.

Blockchain Applications for Climate Solutions

But blockchain, despite its noble claim of democratization of governance, was seen to suffer from its well-noted disharmony with environmental goals, standing accused of producing an astounding 7.7 million transactions generating at least 17.7 Mt of CO2 from Bitcoin and Ethereum's networks in 2017. Thanks to the energy-savings potential from nested transactions and internal program executions that could be made more efficient within encrypted superblocks, this guilt can be removed from the blockchain by enlisting large-scale superconducting infrastructures or routing transactions via sidechains, directed acyclic graphs, or weakly connected supernodes to overcome environmental constraints on the transfer of value across the blockchain.

Evidence of climate change saw a 30-year exponential growth in scientific research with over 13,000 peer-reviewed articles between 2015 and 2017, indicating the urgency of deploying blockchain beyond its nascent tracking function to more sophisticated, smart-contract-based solutions that could effect the necessary reduction in greenhouse gas (GHG) emissions to combat global warming. Together with the surge in climate change research, the Fifth Assessment Report of the Intergovernmental Panel on Climate Change

(IPCC) increased the level of warning to policymakers concerning the staggering costs of global warming, paired with increasing actual global warming trends that reached more than 3 standard deviations over the trends indicated by CMIP5 climate model runs from 1986 to 2017. Similar concerns were echoed by political leaders and economic elites at the World Economic Forum.

CHAPTER 32

Conclusion and Call to Action

In closing, my message is simple yet urgent: The climate clock is ticking away without consideration for our political leaders that may fail to arrive at an effective agreement in Paris, as they did in Copenhagen. The fundamental truth is that the climate threat, now knocking at our door, will soon be harshly upon us. And the future points in only one direction: climate disturbances will become more and more intense. Their impact will be increasingly severe, diffuse, profound, and enduring. We should all take this truth to heart, but make haste: use science and technology along with unconventional conceptions of governance to create climate solutions so that the majority of the world's population never faces government failure.

After a sober examination of the immense difficulty of reaching a global climate consensus, this chapter offered a summary of the heated debate on what is to be done. True to the Hybrid Problem Solving model described in chapter 1, it turns out that a number of scholars and think tanks are examining unconventional and innovative governance structures that seek to internationalize domestic action, and overcome several weaknesses of the United Nations Framework Convention on Climate Change. While they have little in common or are unaware of one another, they share a number of common features, all of which were incorporated in the California – BOG Forum concept, described at the end of the chapter.

Key Takeaways and Recommendations

The politics of compromise place climate engineering measures in the spotlight. Many in the global climate field are investigating possible frameworks for governing technical strategies to combat global warming. The problems outlined in this paper and the suggestions for alternative governance systems raise key questions for these ongoing efforts. In conclusion, the singular difficulty of the governance of climate engineering and the notable ease of the governance of alternative climate solution strategies strongly suggest the superiority of the latter as the focus of the growing policy field. Official research institutions, corporate research, and research conducted by global millionaires are dwarfed by funds available to congest up in the current global climate infrastructure.

Several climate engineering strategies – including CO2 removal from the atmosphere, solar reflectors, and space-based options – could be pursued with much less governance. Debate and analysis about their individual merits and potential dangers are needed, but none presents the particularly immense regulatory challenge of the technique most often discussed today – solar radiation management. Market mechanisms for the governance of technical climate geoengineering raise serious ethical and practical concerns. It is possible to imagine a just and democratic approach to research governance: one in which research is publicly funded, the public sets overall parameters, and those who seek to employ it pay for its indirect costs.

Urgency of Climate Action

Profound strategic questions arise from the sense of urgency associated with the need for action on a decadal time scale and the possibilities for using technologies that seem not to have been invented. Should policy support piecemeal activities that generate knowledge or should it focus on generating system-level robust solutions or real options? How should society mobilize to attack a threat where the adverse environmental impacts may be obvious but the causes are not? Will various parties accept the scientific assessments used to create and justify policies? What combination of regulation, assistance, encouragement, or mandates will nudge people from the stability of the status quo into activities that provide long-term benefits?

Action to combat global climate change would avert harm to future generations. Yet, those who will benefit most from the action - those yet to be born - do not have a voice in today's decisions. They also may be least able to influence today's decisions by voting with their dollars or feet. Resistance to action arises in part from the inherent uncertainty associated with long-term projections and in part from doubts about the willingness and ability of today's society to make changes that its leaders say would confer $25 to $50 trillion in benefits over the next century. People may prefer to remain on a course well established for decades if not centuries without abrupt disruptive change. Unfortunately, failure to act means that the ecosystem services provided by the earth will not be available to everyone.

www.ingramcontent.com/pod-product-compliance
Lightning Source LLC
LaVergne TN
LVHW092050060526
838201LV00047B/1329
9798330391240